Electric Telegraph Series

THE ELECTRIC TELEGRAPH
IN THE UNITED STATES

NUMBER 5 IN THE ELECTRIC TELEGRAPH SERIES

COVER PICTURE
Telegraph operator using a Morse recording instrument
and a Morse finger key.

Alexander Jones

The Electric Telegraph
its rise and progress in the
United States

TGR Renascent Books
2019

PUBLICATION HISTORY

HISTORICAL SKETCH OF THE
ELECTRIC TELEGRAPH
INCLUDING ITS RISE AND PROGRESS IN THE
UNITED STATES

ALEXANDER JONES
1852

This new edited edition published by
TGR Renascent Books
27 Springdale Court
Mickleover, Derby DE3 9SW
United Kingdom
2019

www.renascentbooks.co.uk

All rights reserved.
No part of this publication may be reproduced,
stored in a retrieval system, or transmitted,
in any form or by any means, without the
prior permission in writing of the
publisher, or as expressly
permitted by law.

© Publisher's Copyright: TGR Renascent Books 2019

ISBN 978-1-7987476-4-3

CONTENTS

Chapter   *Page*

    List of Figures iv
    Dedication v
    Publisher's Note vi
    Series Editor's Forward vii
    Author's Preface 1
1   Preliminary remarks 7
2   Chronological statement of discoveries in relation to the Electric Telegraph 11
3   Brief explanation of Terms and Instruments illustrated .. 17
4   Telegraph Litigation—Evidence, Decisions, &c. 33
5   Leading points of contest between the Claims of Morse and Bain 45
6   Local Circuits—relay and receiving magnets 59
7   The Chemical Telegraph 73
8   Dates at which the chief Telegraph Lines in the United States have been built and put into operation 81
9   New Projected Telegraph Lines, to facilitate the transmission of news between the Old and New Worlds, and to unite in communication the Atlantic with the Pacific 87
10 Statistics of Telegraphs in the United States—Plan of erecting lines– Method and Expenses of working them 91
11 Expense of building and operating the lines 101

Contents

Chapter		Page
12	Statistical and other information regarding the operations of several leading lines, supplied by the operators, clerks and superintendents attached to them, in reply to a series of questions	105
	Answers from the O'Reilly Line	106
	Answers from the Bain Line	113
	Answers from the House Line	116
	Answers from the Morse New York and Buffalo Line	118
13	Specimens of Telegraph signal writing and printing	121
14	Connection of the Press with Electric Telegraphs—Earliest communications between New York and other points—Crossing the North River—System of News reporting—Commercial and Congressional ciphers—Telegraphic anecdotes and incidents	125
	Examples from Commercial ciphers	130
	Examples from Congressional ciphers	134
15	Fast methods of telegraph writing—Facsimile transmission of manuscript, printed copy and figures of all kinds	155
	Proposed Electro-Magnetic locks	158
16	Use of the Electric Telegraph in the calculation of Longitude	161
	Municipal telegraphs in cities and towns	165
	Fire alarm whistles	167
	Blasting and submarine explosions by electricity	168
	Electricity in Warfare	170
	Page's Axial Electro-Magnetic Engine	171
17	Foreign Electric Telegraphs—Their rise and progress in Europe—Extent of lines, Plans of construction, and methods of operating them	173

Contents

Chapter		Page
18	Latest accounts of progress and operation of foreign telegraphs	189
	Financial returns of English telegraph lines	189
	Marking Meridian time by electricity	190
	Regulation of time by electric telegraph	190
	Submarine telegraphs	193
	Electric telegraph lines on the Continent	193
	Conclusion	196
19	Charges for telegraph dispatches from New York to all parts of the United States and Canada	197

LIST OF FIGURES

Figure *Page*

    Portrait of Alexander Jones vii
    Facsimile of the 1852 edition title page xi
1 Grove's battery 17
2 Morse registering instrument 21
3 Morse type finger-key 25
4 Bains call apparatus 26
5 House's printing telegraph 27
6 Bain's telegraph 31
7 Signals as arranged by Swaim, Steinheil, Morse, Davy,
    and Bain 56
8 Letters from Steinheil's arranged signs 57
9 Rock blasting by the use of an electric current 170
10 House printing telegraph 172

TO

THE MERCHANTS OF NEW YORK,

IN HUMBLE ADMIRATION OF THEIR PATRIOTISM, INTELLIGENCE, AND ENTERPRISE;

THE FAME OF WHOSE DEEDS ARE RECORDED ON

THE ICY BARRIERS OF ARCTIC SEAS,

IN THE EARLIEST AND LATEST TRIUMPHS OF STEAM,

AND

IN THE UNEQUALLED SPEED OF THEIR SAILING VESSELS,

NO LESS THAN

IN THEIR LIBERAL ENCOURAGEMENT OF WORKS OF INTERNAL IMPROVEMENT,

AND TO WHOSE PATRONAGE,

WITH THAT OF THE PUBLIC PRESS,

THE ELECTRIC TELEGRAPHS ARE LARGELY INDEBTED FOR THEIR SUPPORT AND SUCCESS,

THIS IMPERFECT WORK

IS

RESPECTFULLY INSCRIBED.

PUBLISHER'S NOTE

The international telegram service in Britain, inaugurated by private enterprise in 1845 but soon taken over by the General Post Office (GPO), and latterly by British Telecom (BT), ended in 2003. In the United States the service finished when Western Union sent its last telegram in 2006. As a consequence, most people today, in the age of the Internet, Satellite Communications, Mobile Phones, E-mail, Instant Text Messaging and Fibre-Optic Cable, have no idea that the world was once girdled with thin iron wires strung on poles over thousands of miles of often inhospitable terrain, or that armoured cables lay fathoms deep in abysmal darkness on the bottom of the oceans. It was via these fragile threads that the world once communicated.

The purpose of this Electric Telegraph Series is to publish in new editions some of the many books on telegraphy that first appeared in the Victorian era. Neglected and forgotten, dismissed as no longer relevant, these books are a treasure trove for historians of technology, research students and interested lay persons. The technology and operation of the telegraph very quickly achieved a level of development and sophistication that now seems quite staggering, as a perusal of the books in this series will soon show.

One caveat must be mentioned—the men writing these books were in complete ignorance of the *nature* of electricity, although of course fully conversant with its *effects*. Electricity was often called a "mysterious fluid," by Victorians on the analogy that electricity somehow flowed through a wire like water flows through a pipe. It was not until 1897 that the atom was "split" and J. J. Thomson discovered the electron, the sub-atomic particle that is ultimately responsible for the flow of electricity. It was well into the twentieth century before a coherent theory of electricity was developed and promulgated. Be cautious, therefore, when reading early authors on electricity. Readers who have the need should consult professionals or up-to-date text-books on the subject. For everyone else, the books in this series will provide wonderfully readable and easy to understand accounts of electricity, which while not always strictly accurate according to modern understanding, nevertheless supply everything needed to understand telegraphy, telegraphic circuits, telegraphic instruments and their ubiquitous power sources—hand-turned generators or wet batteries of exceptional size.

SERIES EDITOR'S FORWARD

The author of this book, Alexander Jones, was born on 14 October 1802, in Rowan County, North Carolina, in the United States of America. He was the son of Hardy Jones and the grandson of Samuel Jones, who was of Welsh descent. Hardy Jones was a planter and schoolteacher, who made himself responsible for Alexander's early education. In 1818, at the age of sixteen, Alexander moved to Washington to live with his brother Dabney, and there found employment as a store clerk.

Alexander Jones

Anxious to secure more lucrative employment, the young Jones decided to study medicine and commenced by becoming a pupil of Dr Dunn in Lexington, Kentucky. In 1820 he entered the medical school of the University of Pennsylvania, graduating in 1822. Shortly after, his father died and his estate was divided among seven children. Alexander thought the share allotted to each of his two sisters was insufficient to ensure their proper support and continuing education, so he relinquished his portion to be shared equally between them.

After settling his father's affairs and winding up his estate, Jones and a fellow student sailed from New York for Savannah, Georgia. For the next fifteen years or so he successfully practised medicine in the city, acquiring along the way a sound medical reputation. In 1823 he was made an honorary member of the Georgia Medical Society, and two years later he was named Dean of the Board of Medical Examiners of Georgia. In 1826–27 he served as secretary of the Central Medical Society of Georgia.

During his residence in Savannah, and no doubt influenced by the knowledge that his father had been a planter, Jones developed an interest in cotton. He made several improvements to the cotton gin, which were successfully adopted in the southern states. His foray into this sphere of operations was so successful that in 1840 the British East India Company recruited him (and several other Americans) to help establish cotton growing in India. The men first sailed to England, but while arrangements for their onward journey to India were being made, Jones had a change of heart. It would not, he decided, be patriotic to contribute to the expansion of a crop in India that would compete with an important American southern staple. Accordingly he refused the $6,000 annual salary, plus expenses, offered by the British government. Some of the other men did go to India, but later reports suggested that the project failed because the "Hindoos" did not work well and they observed too many religious holidays to permit proper attention to the cotton fields.

Soon after returning from England, Jones settled in New York and began a new career as a journalist and writer. Under the pseudonym "Sandy Hook", he became a correspondent for the *New York Journal of Commerce* as well as for several British newspapers. This brought him for the first time into professional contact with the electric telegraph. In 1846 he wrote the first news story to be transmitted by telegraph from New York to Washington, which dealt with the launching of the *Albany*, a U.S. sloop of war from the Brooklyn Navy Yard. This experience caused him to grasp the importance of the telegraph in transmitting news reports. However, news-gathering costs quickly became intolerable for individual newspapers, and in an attempt to ameliorate the problem, Jones developed a cipher code that, by 1847, was being used effectively to reduce the cost of transmitting news by telegraph.

It eventually transpired that telegraph companies and unscrupulous operators were monopolising the distribution of news by providing their services only to the highest bidder. In May 1848 David Hale, Jones' employer and editor of the *Journal of Commerce*, arranged a meeting with six other New York papers.

Series Editor's Forward

He proposed that the rival publishers cooperate with the aim of reducing telegraph tolls and securing more reliable news reports from around the nation. Not all the newsmen were enthusiastic at first, but eventually they all recognised that cooperation offered advantages for all. The new group was called the New York Associated Press (NYAP) and Jones, described as "a physician with experience in journalism", was hired as their first agent. At his small office, 10 Wall Street, Jones received dispatches from the telegraph terminal in Jersey City and forwarded them to member newspapers.

One of Jones' first major assignments was the presidential election of 1848. He arranged for news from the Whig convention in Philadelphia to be telegraphed to **Jersey City** where the line, because of the wide expanse of the **Hudson River**, ended. Jones devised a flag system for signalling across the river, so that messages could then be sent on to New York. Unfortunately, the boy assigned to look for the flag saw instead a broker's signal flag and mistakenly notified the papers that **Zachary Taylor** had been nominated. Fortunately for the NYAP, Taylor *was* nominated, but not until the following day.

The success of the NYAP prompted other publishers to form similar organisations, and the Philadelphia Associated Press, the Southern Associated Press, and other companies soon began transmitting news to their member papers. During the **Civil War** years the Western Press Association challenged the dominance of the NYAP, leading to the eventual consolidation of the various cooperatives and the emergence of the modern Associated Press (AP). After three years as the NYAP agent, Jones' resigned in May 1851.

In the following year he published this book, originally with the title *Historical Sketch of the Electric Telegraph, including its rise and progress in the United States*. It gives a detailed account of the history of the telegraph from the discovery of electricity to the organisation of the NYAP. From this time forward until his death, Jones acted as a commercial reporter for the New York *Herald*. Concurrent with his journalistic career, Jones continued to practice medicine. Also, as an outstanding member of the St. David's Society in New York, he pursued his interest in the history and welfare of the Welsh people. He was the author of *The Cymry of '76: Or Welshmen and Their Descendants of the American Revolution*, published in 1855. Jones died in New York on 22 August 1863, after a year's illness. Following a funeral service in St. Alban's Episcopal Church, he was buried in Greenwood Cemetery, Brooklyn.

SAMUEL MORSE AND THE TELEGRAPH IN THE U. S. A.

It has long been received wisdom that Morse was the sole and ingenious inventor of the electric telegraph. In this book Alexander Jones takes considerable pains to establish the truth of this claim, by examining and analysing more than 1000 pages of evidence arising out of the patent trials instituted by Morse and his friends. What he discovered will probably surprise those who uncritically accept the many myths about Morse.

Jones writes that for him to have invented the telegraph, he should have discovered all that was actually discovered by *Galvani, Volta, Oersted, Arago, Ampère, the Davys, Gauss, Faraday, Henry, Steinheil, Wheatstone* and a host of others who had gone before him, embracing nearly half a century. He should have invented the galvanic battery, discovered electro-magnetic motion, discovered the plan of producing electro-magnets, and of varying their power, and adapted them to making signals at a distance. Indeed, he should have discovered nearly all that was known about electricity, and been the first to suggest its application to telegraphing.

Because he did none of these things, he certainly did not invent the art of telegraphy. The distinguished American Professor Henry, in evidence said, "I am not aware that Mr. Morse has ever made a single original discovery in electricity, magnetism, or electro-magnetism, applicable to the invention of the telegraph."

To settle the point, Jones asks, "suppose Morse had never existed, would we have had in operation an electric telegraph in the United States of America?" He answers himself with certainty by replying, "certainly we should."

Gordon Roberts
Derby 2019

Series Editor's Forward

HISTORICAL SKETCH

OF THE

ELECTRIC TELEGRAPH:

INCLUDING ITS

RISE AND PROGRESS IN THE UNITED STATES.

BY

ALEXANDER JONES.

"I'll put a girdle round about the earth in forty minutes." — SHAKSPEARE.

NEW-YORK:
GEORGE P. PUTNAM, 10 PARK PLACE.
M.DCCC.LII.

Facsimile of the 1852 title page.

AUTHOR'S PREFACE

Electricity is the poetry of science; no romance—no tales of fiction excel in wonder its history and achievements. Viewed in its terrible atmospheric manifestations, no element would seem less likely to be brought under the control of man, and, in feebler currents, made to do his bidding, than it: yet, such is the result.

We find it, in one instance, like a skilful chemist, elaborately analysing bodies supposed to be simple alkalis, and showing them to be compounds of metals and oxygen. Again, we find it at work in attempts to manufacture diamonds. Anon, it turns physician, and endeavours not only to heal the sick, but to bring the dead to life. In another case, we find it employed in the plastic art, and, like an expert operator, making beautiful and delicate copies of works of sculpture, and engraving in masses of solid metal. Again, we find it working in the sun's rays, and on the surface of Daguerreotype plates, delineating the human features. It is, again, engaged in dissolving gold and silver, and applying them to the gilding and plating of other metals.

We find it, at another time, employed in blasting rocks from the mountain side, or in removing them from the channels of rivers and harbours. Again, it stands ready to enlist its services beneath the banners of contending armies, to engage on either side, in fearful slaughter and destruction, and then suddenly send to the ends of the earth, the news of its own defeat or victory.

Finally, it turns its electric attention to the movements of "Father Time", and undertakes to give him lessons in regularity and speed. In one instance, we find it conveying messages of intelligence in advance of time over a continent, measuring the degrees of longitude, and dropping copies of its news at each hamlet, village and town, in its flight over mountain peaks:

"Where Alpine solitudes extend;"

across valleys wide, and rivers deep and strong; and as quickly at its post again. Anon, we find it working a hall or city clock, making it accurately mark time in exact seconds, showing its slow but steady flight.

Again, we find it turned policeman; sounding its whistles and alarm bells, to arouse drowsy firemen or sleepy watchmen, calling them quickly to a raging fire, murderous assault, or marauding burglary.

Again, we find its magic power at work in securing the doors and vaults of our buildings, or it is found standing sentinel over our treasures, ready to sound the alarm on the first touch of the robber. It also is prepared to pursue the rogue, fly in advance of his steps, and drop pictures of his person and features at each station on its way. Not only so, but it stands ready to turn coast-guard, to sound whistles or bells over dangerous reefs or rocky shoals, and thus timely warn vessels of impending danger.

Where, in the most extravagant records of fancy—in the wildest dreams of the marvellous—can we find a hero, however lauded and deified, whose most astounding deeds ever compared, for one moment, with the exploits of electricity? Yet, its mighty triumphs are but half revealed, and the vast extent of its extraordinary power but half understood!

In the following pages, we have humbly endeavoured to describe, chiefly, the workings of the electric fluid in reference to electric telegraphs.

Beginning with the earliest notices of electrical laws, exhibited in the practicability of conducting the fluid to a distant point from the place of its generation, whether developed by a frictional machine or a galvanic battery, we have briefly followed it up, through its different epochs, to the present time.

It was not necessary, in our plan, to give a complete history of electricity, nor to notice, in detail, the long array of the names of philosophers made brilliant and immortal by their labours and discoveries, and who contributed so largely to the development of the laws which govern electrical science: such prolixity would fill volumes.

Neither have we been able to bring before the reader the names of all who have, in some way or other, made suggestions, or con-

Author's Preface

tributed, in some form, to the establishment of electric telegraphs.

Atmospheric electricity has undoubtedly coexisted with other elements of creation. And some have supposed that it was the primary element, employed in the fiat of creation, and yet remains that universal power of attraction and repulsion, by which worlds are sustained in their orbits; while, at the same time, it is the life generating and supporting principle in all existing forms of vitality.

Thales, the oldest of the seven wise men of Greece, taught that water was the primeval element of all other things appertaining to the earth, and this doctrine was current up to the period of Paracelsus, about the dawn of the sixteenth century, when the one element was extended to four—water, air, earth, and fire. These elements were soon after found, themselves, to be compound.

It seems to us, that it would have been wiser, had electricity been fixed upon as the simple, all-powerful and pervading element, instead of water.

We derive the word *electricity* from the Greeks, who discovered that when amber (called by them *electron*) was rubbed, it exhibited properties of attraction which it did not otherwise possess.

This property in *amber*, it is said, was first observed by Thales, six hundred years before the birth of Christ. Several accounts of electrical phenomena were also recorded by Aristotle, Theophrastus, Pliny, Caesar, and Plutarch.

It was not until the seventeenth century, that the attention of philosophers was strongly attracted to the subject of electricity, and various new facts relating to it were first discovered. Early in the eighteenth century, it received increased attention, and many of the greatest minds were led to investigate its laws, and to ascertain the nature of its effects, which, on being published, surprised the world by their striking novelty.

In 1745, the German experimenter Ewald von Kleist, a philosopher of Leyden, discovered (what has since borne, not his name, but that of the town in which he lived) the *Leyden jar*, or *phial*. This gave a new impulse to electrical science, which was soon after followed by the discoveries of Dr Franklin and others, some of which we have alluded to in our brief chronological statement. M. Monnier, the younger, discharged a Leyden jar through a wire of four thousand feet in length, but could not estimate the velocity of its speed.

Elliott, of Edinburgh, first constructed an electrometer for measuring the quantity or force of the fluid. Abbé Nollét, of France, also showed by experiment that the electric fluid could be conveyed to considerable distances.

It is said that the Marquis of Worcester alludes to telegraphs in his famous *Century of Inventions.* Robert Hooke, in 1684, presented a paper to the Royal Society, "showing a way how to communicate one's mind at great distances", probably by a visual telegraph.

In 1773, Odin, of France, suggested the possibility of instantaneous communication.

For discoveries and suggestions made by other parties in reference to electricity and its application to telegraphs, we refer to our chronological table.

In the present age, when education is becoming so universally diffused, when the knowledge of great and important sciences has passed from the few to the many, when new discoveries are rapidly following each other in all the useful pursuits of man, encouraged in our happy country by free political institutions, we must believe that new discoveries will be made in electricity and its applications, and that, in the prospective and brilliant future, the perfection and extension of electric telegraphs will meet with their due share of success.

In submitting our brief and imperfect production to the public, we feel sensibly impressed with the diffidence common to authors, and especially to those unused to appear before the world in that capacity. We have been compelled to discuss subjects of painful delicacy, either in defence of truth or in the way of personal explanation. In doing this, we have endeavoured to respect the feelings and merits of all interested parties, and neither to give nor intend personal offence to any person whatever.

We have had to write our work at leisure moments, snatched from other daily occupation, and chiefly during evenings, or at night. We therefore have to crave the indulgence of the reader for such errors as may here and there present themselves. Our aim has been, to describe electricity and electric telegraphs in such a

Author's Preface

popular form, as to enable a party not previously and fully initiated, to gain a clearer and better knowledge of the subject.

The facts regarding electric telegraphs have multiplied so rapidly of late years, and their extension has become of such general and public utility, that some work of the kind, embracing all the leading points of their history up to the present period, seemed necessary.

Trusting that our labours will be found, to some extent, both useful and entertaining, we have the honour to be:

The public's very obedient and humble servant,

THE AUTHOR

April 1852.

1
PRELIMINARY REMARKS

It has been supposed by many that the idea of an electric telegraph did not exist anterior to Dr Franklin's experiments made to test the identity of the lightning with that generated by an electrical machine; but such was not the fact. Grey and Wheeler, of England, as early as 1728, showed that electricity could be conducted to a great distance. Dr Watson, of England, was the first to propose the construction of an electrical telegraph, in 1747. Dr Franklin's attention was first drawn to the subject of Electricity between 1745 and 1746. It was not, however, until June 1752, that he raised his kite and drew an electric spark from a passing thunder cloud, the result of which he communicated in a letter to Mr. Collison, of London, dated Philadelphia, October 19, 1752. The Royal Philosophical Society of London, it is said, were so incredulous regarding the reality or value of his experiments, that they refused to admit them to record in their *Transactions*. Be this as it may, Mr. Collison soon after published Franklin's experiments in a pamphlet form, which was translated into several languages, and attracted almost universal attention. There are many curious things connected with the progress of electrical telegraphs, from the earliest suggestions regarding their practicability up to their latest improvements and application, as we at present see them in use in this country and in Europe. The introduction of electrical telegraphs, and their daily employment in the speedy transmission of news, forms one of the most remarkable eras of the nineteenth century, and marks strongly and indelibly a stage in the conquest of mind over matter.

Such is the importance of the subject, that every link in its history should be clearly noted and recorded. Like most other great discoveries, its commencement, rise, and progress do not belong to, nor have they originated with, a single individual. But, from its earliest conception to the present time, its utility has been rendered available, step by step, by the labours and discoveries of a large number of distinguished scientific men, whose developments in electricity, electro-magnetism, and permanent batteries, &c., were all given to the world in scientific publications, at different periods of time.

The mass of mankind, as yet, do not seem to realise the vast consequences to which the use of electric telegraphs may lead. They do not, and cannot, comprehend the future in the applications of electricity. It is difficult for anyone to give a clear and popular idea of the manner in which electric telegraphs are really worked—how it is that a message of more or less length is so instantaneously, as it were, reproduced, at the distance of four or five hundred miles, copies of which are dropped on the way. The public see posts erected, and wires stretched on them. They also see machines of small dimensions, in telegraph offices, managed by operators, and may imagine what a battery is that generates electricity, as well as that a piece of cold iron may be converted into a magnet, while the electricity is passing round it through a coil of copper wire, which ceases to be a magnet, or to attract, when the current is cut off. But there are few who can comprehend, after all, exactly how the thing is done—how one system of telegraph differs from another—why Morse's machines make blank dots or indentations, on white slips of paper—how House's prints messages in Roman letters—and how Bain's employs electricity as a chemical agent in the discoloration of paper, by which messages are transmitted. And unless parties to whom the whole is to be explained possess some previous knowledge of electricity and its laws, it is difficult for them to understand it. Neither can the peculiarities of the various telegraph machines employed here and in Europe, be explained without diagrams. It is easy to understand the steam engine, the workings of a steamboat, cotton, or other more palpable or demonstrative machinery; but not so the workings of the telegraph machines, or the nature of the subtle fluid by the agency of which they are actuated. It would extend our narrative beyond all reasonable limits were we to attempt to give any descriptive notice of the different machines in use, which now amount to more than a dozen, here and abroad.

Preliminary Remarks

Of one thing all may feel assured—that the electric telegraphs are yet in their infancy. The time must come when they will work a great revolution in the affairs of men—in their social, political, and commercial intercourse. The time will come when all the proceedings in Congress will be transmitted, *in extenso*, to all parts of the Union daily—when they will become the medium of communication for all letters of consequence, passing between distant points of the Union, instead of their slow transportation by mail. The time will come when New Orleans, the city of Mexico, San Francisco, and Astoria, on the Pacific, will be in as constant, steady, and daily communication with New York, as Albany, Philadelphia, and Boston. And furthermore, the time must and will arrive, be it fifty or a hundred and fifty years hence, when great telegraph lines will unite all parts of the civilised world in daily communication.

When the telegraph is well established hence to the Pacific, it will be *practicable*, by proper means and energy, for it to be extended northwest to Behring's Straits, which is only about thirty miles wide, and found by Capt. Cook to be only six fathoms in depth. Once carried into Asia, across these Straits, by the liberal co-operation of the Emperor of Russia, they could be carried to St. Petersburg, and from thence to Germany, where, intersecting with the present lines between London and Paris, and other European capitals, they would be put into communication with New York.

All idea of connecting Europe with America, by lines extending directly across the Atlantic, *is utterly impracticable and absurd*. It is found on land, when sending messages over a circuit of only four or five hundred miles, necessary to have relays of batteries and magnets to keep up, or to renew, the current and its action. How is this to be done in the ocean, for a distance of three thousand miles? But, by the way of Behring's Straits the whole thing is practicable, and its ultimate accomplishment is only a question of time, made near or remote by the progress of population and civilisation.†

---

† Editor's Note: Only six years after Jones wrote this paragraph, a cable was successfully laid on the bed of the Atlantic Ocean, joining America to Great Britain. Although it failed after a short time (owing to mismanagement), two more cables were laid and were in operation by 1866, and no less than fifteen cables crossed

The time must come when the political movements and discussions, the state and condition of trade and commerce, the ups and downs of daily life, will be daily and simultaneously published in Paris, Vienna, London, and New York, thereby ushering in a higher degree and wider sphere of civilisation, and of "peace on earth and good will to man".

---

the Atlantic by 1894. Conversely, a line by way of the Behring Straits (today usually spelt Bering Strait) proved not to be practical after the failure of the Collin's Overland Expedition (financed by Western Union) to construct a telegraph line through the icy wastes and forests of northern Canada and Alaska.

2
CHRONOLOGICAL STATEMENT
OF DISCOVERIES IN RELATION TO THE
ELECTRIC TELEGRAPH

Early in the eighteenth century, when it was discovered that electricity generated by a frictional machine could be conveyed to a long distance, the idea of applying it to telegraph purposes was suggested, and various attempts made to put it into practical use.

The efforts made to establish electric telegraphs divide themselves into four periods:

First, from the development of electricity by friction, to the discovery of galvanism, or the production of electricity by the chemical action of acids upon metals, in 1790 by Galvani, and by Volta in 1800.

Second, from 1790 to 1820, when Oersted discovered electromagnetism, and Ampère showed its applicability to telegraph purposes.

Third, from 1821 to 1831, when Professor Joseph Henry discovered the mode of constructing improved magnets, in connection with properly arranged batteries, so as to produce mechanical effects at a distance.

Fourth, from 1832 until the present time, 1852.

First Period

1726 Wood, of England, discovered that the electric fluid could be conveyed a long distance, by conducting wires.

1746 Winkler, of Leipsic (Leipzig), discharged a Leyden jar by a friction machine, through a wire of considerable length; and on that occasion the River Pleis (River Pleiße) formed part of his circuit. (*Priestly's History of Electricity*, p. 59.)

1747 Dr Watson, of England, extended the experiments over a space of four miles, at Shooter's Hill, near London, comprising his circuit of two miles of wire and an equal distance of ground. He is believed to have been the first who suggested the application of electricity to telegraph purposes. (*Phil. Trans.*, vol. xiv., 1848.)

1748   Dr Franklin set fire to spirits by an electric current sent across the River Schuylkill on a wire, and allowed to return by the river and earth. (*Sparks' Life of Franklin*, vol. v., p. 210: Boston.)

1784   Lomond, of France, communicated telegraph signals to a neighbouring room, by means of a pith ball electrometer, acted upon by electricity. (*Young's Travels in France*, 1784, vol. ix., p. 79.)

♦   Reiser illuminated letters upon plate glass, formed of tin foil by means of electricity. (*Voight's Magazine*, vol. i., part 1st.)

1795   Cavalo proposed to form an electric telegraph by firing a gas pistol at the distant end of a wire, and thus to give signals. (See his *Treatise on Electricity*, vol. iii., p. 295.)

♦   M. Savary attributes the first idea of an electric telegraph to Dr Franklin. (See *Amoyt's Memoire Comptes Bendus French Academy*, sitting July 1838, pp. 80, 81, 82.)

1798   Betancourt established a telegraph between Madrid and Aranguez, twenty-six miles, through which a current of electricity was passed, and gave signals for letters. (Given in the work of *Gauss and Webber*, on the authority of *Humboldt*.) "He constructed an electric telegraph in Spain, by which the infanta, who saw it operate under his own eyes, was specially informed of a piece of news from a very great distance". (See *Amoyt's Memoire* above.)

Second Period.

1809.   Sömmering constructed the first galvanic telegraph at Munich, which operated by the decomposition of water; and which he also caused to ring a bell at the opposite end of the wire. (*History of Electric Telegraphy*, by Abbé Moigno, p. 300. Paris, 1849.)

♦   Sömmering speaks of the telegraph of Reiser and Salva, which had preceded him; and also of the experiments which had laid the foundation of the telegraph, by Le Mounier and Watson.

♦   Sömmering's telegraph was the first decomposing or chemical telegraph; and can be even now successfully, but less rapidly, worked than Bain's.

1816 Dr John Redman Coxe, of Philadelphia, proposed to establish an electric telegraph, and to make signals at a distance by the decomposition of water and metallic salts, causing a change in colour to ensue. (*Thompson's Annals of Philosophy*, vol. vii., p. 162. *Journal Franklin Institute*, vol. xx., p. 325. Philadelphia, 1837.)

1823 Francis Ronalds, of England, proposed a telegraph by the use of frictional electricity. In his arrangement there were clocks at the stations which kept time with each other, and which were furnished with a light disc of ciphers in place of hands, having twenty different signs towards their circumference. At the moment the proper sign on the disc passed before the index at one station the spark was discharged, and an electrometer placed at the other discharged and caused the sign on the disc at the other to be noted. This telegraph is stated to have extended to Hammersmith, eight miles, and to have used the discharge of a gas pistol as an alarm. (*Encyclopaedia Britannica*, 1842, 7th edition, Letter E, p. 662. *Steinheil's Annales of Electricity*, vol. iii., p. 446.)

THIRD PERIOD, 1820 TO 1831

1819 In this year Professor Oersted, of Copenhagen, discovered electro-magnetism, or electro-magnetic motion. (See *Annals of Philosophy*, vol. xvi., p. 273.)

1820 Ampère, of France, discovered the electro-magnetic telegraph. This he constructed of as many wires as there were letters, and used the deflection of the needle as a signal. He broke and renewed the circuits by finger-keys, something similar to those of the keys of a pianoforte. (*Annales de Chemie et de Physique*, 1820, vol xv, p. 75. *Expose des Nouvelles Decouverts par Ampère et Babinet*, Paris, 1822. Amyot, *Comptes Rendus*, 1838, p. 81.)

1825 Barlow, of Greenwich, England, attempted to put a galvanic telegraph in operation in this year, but was thwarted by the diminution of the fluid, when he endeavoured to transmit it for a great distance, so as to produce mechanical effects. This difficulty the discoveries of Henry, however, afterwards overcame. (See *Testimony for Defence in case of F. vs. R.*, p. 250.)

♦ Mr. Sturgeon, of England, in the same year constructed the first *electro-magnet*, by coiling a copper wire round a piece of iron, of a horse shoe form, the bent turns of the wire being so far apart as to prevent contact. He found that when the electric fluid passed through this coil, the enclosed iron became a magnet, and was again demagnetised on breaking the current. The wires were afterwards coated with non-conducting substances, and wrapped around the iron in close contact, as we now see them. An account of his experiments was first published in Nov. 1825, in the *Transactions of the Society for the Encouragement of Arts in England*. (See also *Annals of Philosophy* for Nov 1826.)

1826 Harrison Gray Dyar erected a telegraph on Long Island, in New York. He used frictional electricity and dyed marks on chemically prepared paper by the passage of sparks. (See *Evidence for Defence, French* vs. *Rogers*.)

1831 Professor Joseph Henry, then of Princeton College, discovered a method of forming magnets of intensity and of quantity, produced from correspondent batteries, and by the use of which, with relay magnets, &c., proposed by him, he made known the practicability to produce mechanical effects at a great distance; without which no mechanical electro-magnetic telegraph could ever have been put in successful operation for great distances, say from 1000 to 2000 miles. (See 19th vol. of *Silliman's Journal of Arts and Sciences*; also *Henry's Evidence for Defence, F.* vs. *R.*, p. 253.)

FOURTH PERIOD, 1832 TO THE PRESENT (1852)

1832 Baron Schilling, of St. Petersburg, contrived a deflective magnetic telegraph. This telegraph had an alarm bell connected with it. It was a step in advance of previous electro-magnetic telegraphs. (See *Sturgeon's Annals of Electricity*, vol. iii., p. 448, 1839.)

1833 Gauss and Weber first constructed the simplified electro-magnetic telegraph. It was Gauss who first employed the incitement of induction, and who demonstrated that the appropriate combination of a limited number of signs, is all that is required for the transmission of communications. Weber discovered that a copper wire 7,400 feet long, which he carried over the houses and church steeples of Gottingen, from the Observatory to the Cabinet of Natural Philosophy, required no special insulation. This was an important point of discovery in the construction of telegraph lines, and it is made available to the present time. (*Sturgeon's Annals of Electricity*, vol. iii., p. 448, 1839; *Gott. Gel. Am.*, p. 1273, and *Schumacher's Jahrbuck*, 1837, p. 38.)

1837 Steinheil constructed and put in use in July of this year his *Registering Electro-Magnetic Telegraph* between Munich and Bogenhausen. By the deflection of a needle he produced dots, or short marks, on fillets of paper to stand as signals for letters, &c., the paper being drawn forward by clockwork, in an endless slip or ribbon. (See *Sturgeon's Annals of Electricity* vol. iii. p. 449-450. — *Comptes Rendus of the French Academy*, sitting September 10th 1838 pp. 590-593.)

1837 (June 12th), the deflective Electro-Magnetic Telegraph of Cooke and Wheatstone was patented in England. They first employed receiving and relay magnets. (*London Repertory of Patent Inventions*, 1839, vol. xi., new series p. 1.)

♦ In October 1837, Samuel F. B. Morse of New York entered his first caveat for an "American Electro-Magnetic Telegraph", in which he chiefly relied on a kind of type and port rule for making signals by the mechanical force of Electro-Magnetic Motion.

In his letter to the Secretary of the Treasury dated Dec. 6th 1837, he stated that he had devised but not tested his improvement, until a few weeks previous to the 27th September 1837. Morse claims that he first thought of a Magnetic Telegraph when coming to the United States in the ship *Sully*, in 1832. Professor Charles T. Jackson of Boston claims the credit of having suggested the first idea of an Electrical Telegraph to Professor Morse, while coming over in the same vessel as a fellow-passenger. But neither appeared to have been aware, at the time, of all that had been previously accomplished before them. Morse obtained a patent in France in 1838; and in the United States in 1840. His various re-issues, new patents, &c., we have subsequently alluded to under another head.

1838 Edward Davy of London had his patent sealed for a chemical Telegraph, which was enrolled January 4th, 1839. In his plan he employed chemically prepared paper, similar in its general character to that employed in the instrument of Bain. (See his *Specification in Evidence for Defence*, French *vs* Rogers, pp. 34-44.)

1843 Alexander Bain obtained his English patent for *Electro-Magnetic Clocks*.

1846 Mr. Bain obtained his English patent for his improved *Electro-Chemical Telegraph*.

1848 Mr. Bain entered his claim for an American patent, which was confirmed to him by Judge Cranch in 1849. (See *Specifications*, pp. 146-157, also 100, 112, and 115, *Evidence for Defence*, French *vs* Rogers.)

1848 Royal E. House of New York obtained his patent for his very ingenious and valuable Printing Electric Telegraph.

1848 Zook and Barnes of Cincinnati invented a modification of the Electro-Magnetic Telegraph, by combining fixed magnets with the use of electro-magnets.

1849 About this time Mr. Horn of New York invented his Igniting Telegraph, which made dots and lines by burning them in slips of revolving paper by the heat of the electric fluid, while passing.

About the same time, a Mr. Johnson, of New York, contrived a machine worked by Electro-Magnetism to let shot drop on to slips of paper, which, being pressed at the same moment, left visible marks which stood as signs for letters.

An Axial Telegraph was also about the same time proposed by Daniel Davis, of Boston, which with that of Horn's and Johnson's have not come into any general use.

3

BRIEF EXPLANATION OF TERMS
AND OF INSTRUMENTS, ILLUSTRATED

To impress the mind of the uninitiated reader with a clearer idea of the meaning of certain terms employed in the discussion of the subject, we will endeavour to explain them.

1. There are two well-known modes of developing electricity. First, by friction, such as is used in a common electrical machine, second, by the solution or decomposition of metals in acids—two metals being usually employed, of opposite conducting qualities, such as zinc and platina (an alloy of copper and zinc), immersed in diluted sulphuric or nitric acids.
2. Electricity produced by friction is called "Frictional Electricity".
3. Electricity produced by the immersion of metals in acids is called "Galvanism". And the arrangement of the metallic plates, the zinc and platina alternating (or copper instead of platina may be used), in a row, one end terminating in a zinc, and the other in a platina or copper plate, with the cups of acid for their immersion, is called a *Battery* (Figure 1). A battery is formed thus:- Take any number of glass cups, in the form of

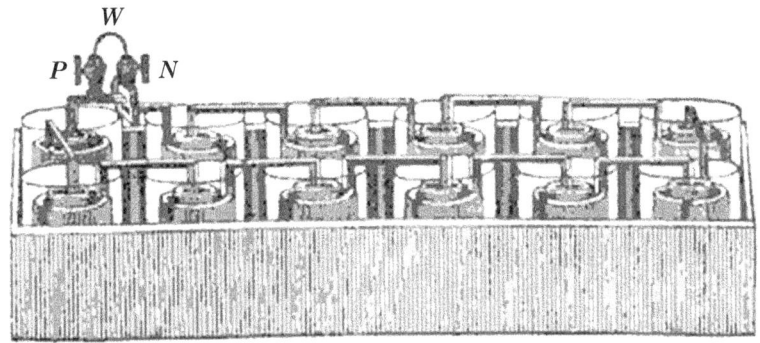

Fig. 1
A battery formed by twelve cups on the plan of Mr Grove
P positive pole *N* negative pole *W* wire

large sized glass tumblers, say, from twelve to one hundred. Take an equal number of thick cast zinc cups, half an inch less in diameter than the glass cups, open at bottom, and with an open slit on one side. Set a zinc cup thus formed in each glass tumbler. Again, inside the zinc-shell, place a small porcelain cup, unglazed, and closed at the bottom. Into this porcelain cup place a thin piece of platina foil, half to an inch wide, and two or three inches long. The upper end of this platina foil is connected with the zinc cup in the next adjoining tumbler, by a slip of metal, and the platina in that cup to the zinc in the next, and so on through the whole series. The cups may be set in a single or double row on a table, or they may be placed in a circle. The arrangement begins with the platina foil at one end, to which a wire is attached, called the positive pole, and terminates with a zinc cup at the other end, to which a wire is also attached, which is called a negative pole. To bring a battery thus formed into action, sulphuric acid (oil vitriol), largely diluted with water, is poured into the glass cups, so as to fill all the space inside and outside of the open zinc cups, while the porcelain cups are filled with diluted nitric acid (*aqua fortis*), and surround the platina foil. The weak sulphuric acid in the outer chambers may also be rendered more active, by the addition of a slight portion of nitric acid. When thus prepared, electricity is actively developed at the poles or ends of the opposite wires, whether long or short. To render the decomposition of the zinc cups less rapid, they are usually coated with quicksilver, merely by rubbing it over their surface. When the battery is not in use, the metallic portions are removed from the acids in the glass and porcelain cups.

4. The platina plate at one end of the row, is called the *Positive Pole*, and the zinc at the other, the *Negative Pole*.

5. A copper wire soldered to the negative pole, or zinc plate, and another to the platina, or positive pole, however long, are called *Circuits*. And when their ends are brought together, a shock or spark of electricity will be produced. It is supposed that there is such an action of the acids on the metals, as to cause an accumulation of the fluid at the platina end of the

Brief Explanation of Terms and of Instruments Illustrated

battery, and a diminution at the zinc end; and hence there is a strong tendency in the fluid to gain an equilibrium, by reaching the zinc pole by the best conductor, and hence equalise or reduce the supply at the platina end.

6. A *Current* is the passage of the electric fluid through a wire, from one pole to the other. To make the subject still more plain, suppose a house, set due north and south, to represent a *battery*. Suppose its north end to represent the positive pole, or the platina end, and the south end the zinc pole. Suppose a wire to proceed from the north end to distant villages and towns, a hundred miles off, and suppose another wire from the zinc end proceeds by another route, and is brought near to it, both being secured on poles, and supported by glass, to prevent the fluid reaching the ground. Now a man, standing at the ends of the two wires, can produce an electrical spark between them by merely bringing them in contact. When the two ends are in contact, the *circuit* is said to be closed and the *current* passes silently and instantly over a space of 200 miles, from one end of the house to reach the other. But, if the wire from the north end of the house is made to terminate a hundred miles off, and then connected with a piece of metal stuck in the ground, the current takes the ground as a conductor, and reaches the zinc at the south end of the house; hence, when the ground is used in place of one of the wires, which is the case in all telegraphs, it is called the *Ground Circuit*, which was discovered, in its application to telegraphs, in 1837, by Steinheil, of Germany. This fact enables news to be forwarded and answers returned, by using a single wire instead of two.

7. *Electro-Magnetism.* If a piece of iron wire, or a common needle, be laid on iron filings, and the ends of the two wires from each end of the house or battery be attached to its respective ends, or one end of the needle with the ground wire or plate, and the other with one wire of the battery, it is found while the electric fluid is passing through it, to become a magnet and to attract the iron filings. The moment, however, the passage of the fluid or current is interrupted, by removing the end of either wire, the piece of iron or needle loses its magnetism and the filings

The Electric Telegraph in the United States

drop off.
8. *Electro-Magnets.* It is found, if the iron is wrapped a great many times by a coil of covered copper wire, having its ends free, and that if the ends of the wires of the battery are attached to them, and the current or fluid made to pass through the copper coil instead of the metal, that the enclosed metal, while the fluid is passing, is converted into a powerful magnet, but ceases to be such on the interruption of the current by breaking the connection. Thus the iron may be alternately magnetised and demagnetised, at the will of the operator. The Electro-Magnetic Telegraphs are all founded upon this simple fact. Because, by having a small piece of metal so poised that a small spring separates its end from the iron when no electricity is passing, yet when the current is put on, the iron is instantaneously sufficiently magnetised to attract the free end of the lever against itself, by overcoming the power of the spring, but releases its hold again the moment it loses its magnetism, and the spring again throws it off. Now if the end of the lever has a sharp point, and a ribbon-formed slip of paper is reeled off by a train of clockwork between the end of the lever and the magnet, it is evident that every time the lever is attracted to the magnet it will indent the paper; and these indentations, by varying their numbers, may be made to represent the letters of the alphabet. Such is, in substance, Morse's Telegraph. Here (Figure 2), we have a large spool *S*, on which the strip of paper is wound, and clockwork with rollers, which give the strip a steady motion onwards under the stylus upon the lever of the electro-magnet *D*. There is a stop motion of the lever used to draw attention, by which the clockwork is brought to rest in a few seconds after the lever ceases to act, and which is released again by the first motion of the lever. This ingenious mechanical combination of Morse forms a simple and efficient registering machine.

To increase the mechanical force of the lever and magnet, let us suppose that at the station a hundred miles off, a small house (or battery) is built similar to the large house, say with not over one-fortieth of its number of rooms (cups), in a house

Brief Explanation of Terms and of Instruments Illustrated

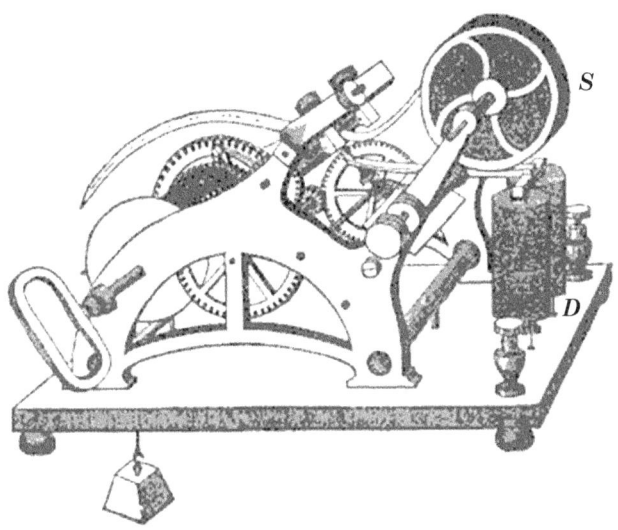

Fig. 2
Morse registering instrument

of one hundred rooms, and that a wire from its north end, of only forty feet long, is made to convey a current to increase the power of the local magnetic force or mechanical action of the fluid conveyed one hundred miles from the big house. This little house is called a *local battery*, and the wire attached to it a *local circuit*. Mr. Bain, to increase the metallic decomposition and chemical action in his Telegraph, divides or sets off a few rooms (cups) at one end of his big house, and conveys a wire from them, to change by its silent chemical action (not mechanical, for he can use a weaker current than Morse)—the colour of the prepared paper on a revolving metallic disc, the signs being made by breaking and renewing the circuit. The Chemical Telegraph was discovered by Edward Davy, of London, and improved by Bain. Bain calls his arrangement, which is a part of the big house and main line, a *Branch Circuit*. We learn that so slight a current of electricity is necessary to produce chemical action in the Bain instrument, that even the use of a branch circuit has been dispensed with on the Bain's New York and Boston line. Effects can be chemically produced, or messages recorded by Bain's Instrument, through the action of a current which would be too feeble to excite the mechanical action of Morse's instrument. In this respect it has an advantage over mere mechanical telegraphs.

9. *Combination of Circuits. Relay and Receiving Magnets.*—It is impossible to transmit a current from a single battery in New York to New Orleans, over an uninterrupted wire, unless the current should acquire from some source an augmentation of fluid on the way. Just as it becomes necessary to have canal feeders, to supply the water lost by absorption, or evaporation, it was found that if the length of the wire be extended several hundred miles from the house, instead of one hundred, that the current in its transmission grows weaker, and tends to lose its force by diffusion, particularly if the air be damp, or a rain storm should prevail. To supply an additional amount .of electricity on a long line, batteries (or houses) are located on an average of every hundred miles, chiefly at intermediate telegraph stations, in cities, or villages. Relay batteries and receiving magnets were first discovered by Professor Henry and used by Cooke and Wheatstone.

To understand how, at the termination of a wire one hundred miles from the battery or starting point, an operator can bring into action another current generated at a way-house (battery), on to another wire of another hundred miles (more or less), let us suppose the following arrangements:- The way-battery is stationed near the termini of both wires, and is supposed to be in connection with the second line of wire. Suppose a man (an electro-magnet) is made to stand between the ends of the wires, at every hundred miles. In his left hand he holds the end of the first wire, while his right arm is extended in an opposite direction, and the knuckles of the hand made to come almost in touching distance of the end of the second wire, which has affixed to it a spiral spring, attached to a small piece of platina. When the operator at the starting point closes the circuit, or puts on the electric fluid, the piece of platina is drawn against the knuckles of the hand, which at the same time forms the connection through the whole distance, and with the local batteries; the quota of fluid generated by each, being at the

Brief Explanation of Terms and of Instruments Illustrated

same moment brought into action on the entire line. In place of one man it has been proposed to employ two (two electro-magnets) between the termini of wires at each hundred miles, and to bring them into contact, by a mechanical contrivance something after the fashion of two men shaking hands, while each holds the end of a wire in the other hand.

Mr. Charles Bulkly, superintendent of the Washington and New Orleans telegraph line, has contrived a modification of the relay or connecting magnets, for bringing the main lines into unison of action, which is said to work very well. To comprehend more fully the nature of the arrangements at present most commonly in use, we subjoin the following from a description of instruments published by Daniel Davis, of Boston, 1851, pp. 21-23, which describes the arrangements for bringing a combination of circuits into action more in detail, and which are most commonly used on telegraph lines at the present time.

The effect of the combination of circuits is to enable a weak or exhausted current to bring into action, and substitute for itself, a fresh and powerful one. This is an essential condition to obtaining useful mechanical results from electricity itself, where a long circuit of conductors is used, and accordingly it received the attention of early experimenters with the telegraph. This principle seems to have been first successfully applied by Professor Joseph Henry, of Princeton College, in the latter part of 1836. He was thus enabled to ring large bells at a distance, by means of a combined telegraphic and local circuit. In the early part of 1837, Wheatstone in England,† used a combining instrument, which consisted of a magnetic needle, so arranged as to dip an arch of wire into two mercury cups, when deflected by a feeble current, thus completing the circuit of a local battery, which struck a signal-bell. Davy patented in England, in 1838,‡ a system of combined circuits, for four different purposes connected with his telegraph.

He brought into action a local circuit, 1st, to discolour or

---

† *London Repertory of Patent Inventions*, 1839, vol. xi.
‡ *London Repertory of Patent Inventions*, 1839, vol. xii.

dye, by electro-decomposition, the calico on which he registered his signs; 2nd, to actuate an electro-magnet regulating the motion of the calico; 3rd, to direct the long or telegraphic circuit to either of two branches, by means of a receiving instrument placed at their point of meeting, and operated upon, from a distance; 4th, he provided for a complete system of relays of long circuits. His instrument resembled Wheatstone's, only the contact was made by two surfaces of metal, without the use of mercury.

The receiving magnet used by Professor Morse is a very slight modification of his register, the platina point for completing the local circuit being substituted for the marking point. The magnet is surrounded with helices of fine wire, which multiply the effects of the feeble current, and the whole instrument is constructed with delicacy. By Morse's patent of 1840, this is applied to the combination of long circuits, or the *relay* of currents; and by his patent of 1846, it is applied to operating the register by a local or office circuit. The electro-magnet, armature and lever, constituting the chief part of both these instruments, is simply the electro-magnet of Professor Henry, described in 1831.

In a line of telegraph of several hundred or thousand miles, any number of receiving magnets may be interspersed, as they do not interrupt the circuit. Each one of these may work a local register, and thus the same message may be recorded at a multitude of places, practically at the same moment of time. If the receiving magnet is to effect a relay of currents, the motion of its lever brings into action a powerful battery on the spot, which works the next receiving magnet in succession, and so on.

The use of the receiving magnet, however, for the purpose of *relay* of the galvanic force, may be dispensed with by simply increasing the number of cups, and distributing them in groups along the line. Thus Mr. Sears C. Walker, of the Coast Survey, writes, "We have made abundant experiments on the line from Philadelphia to Louisville, a distance in the air of nine hundred miles, and in a circuit of eighteen hundred miles. The performance of this long line was better than that

Brief Explanation of Terms and of Instruments Illustrated

of any of the shorter lines has hitherto been. I learn, from an authentic source, that the same success attends the work from Philadelphia to St. Louis, A DISTANCE IN CIRCUIT OF ONE TWELFTH OF THE EARTH'S CIRCUMFERENCE. The number of Grove's pint cups used is about one for every twenty miles. It is natural to conclude, from this experiment, that, if a telegraph line round the earth were practicable, *twelve hundred* Grove's pint cups, in equidistant groups of fifties, would suffice for the galvanic power for the whole line." (See *Silliman's Journal*, March 1849.)

By the foregoing explanations, we are better enabled to understand what is meant by *relay batteries*, *relay* or *local magnets*, *receiving magnets* and *combined circuits*.

If a registering instrument be placed at each station in connection with the way-magnet, it is clear that it can be operated without arresting the current, and hence copies of the same message can be read off at each intermediate station between points however distant. This is called dropping copies. This facility of having messages dropped at intermediate stations, was said to have been first demonstrated in this country by a Baltimore operator. The plan at first was to receive a message at one station, and then re-forward it, and each way-operator had to be on the look-out. This party finding this irksome for through messages, left his instrument in the circuit, when he found he could take a copy of what was passing without interfering with its transmission to its ultimate destination.

10. A *Finger-Key* is merely a small office spring-lever (Figure 3), with an ivory button at one end, on which an operator presses

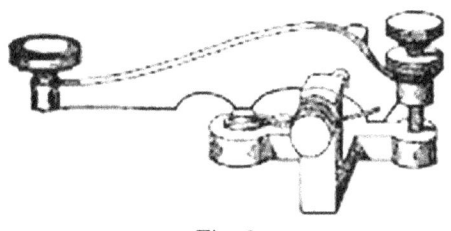

Fig. 3
Morse type finger-key

his finger when he wishes to connect the circuit, or secure the passage of the fluid, and raises it when the spring throws back the lever and breaks the circuit—which can be done with great rapidity. This may be understood from the fact, that it requires from one to four or five motions of the finger to make a single letter or figure; and yet an expert operator can send at least 80 to 100 letters per minute, and about 1000 words or more per hour. Further improvements in time, must develop new and important powers in electric telegraphs.

11. *The Call.*—As the Bain Electro-Chemical Recording Telegraph acts silently, it became necessary for him to contrive this *call* whereby one operator at a distant point can give notice at the opposite end of the wire that he is ready to send a message; whereupon he can proceed to put the chemical recording instrument into connection with the current ready for operation.

The *call*, commonly used on the Bain lines, is represented in Figure 4. It consists of a U-shaped receiving magnet, placed horizontally on the board, with two coils of wire surrounding the legs. An armature, supported on an upright bar, so as to form a cross, is seen in the figure, before the poles of the magnet.

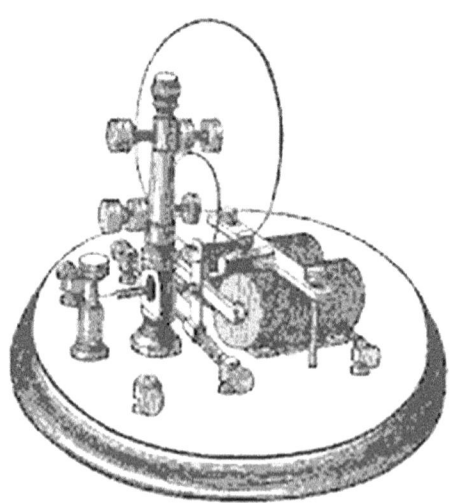

Fig. 4
Bain's call apparatus

Brief Explanation of Terms and of Instruments Illustrated

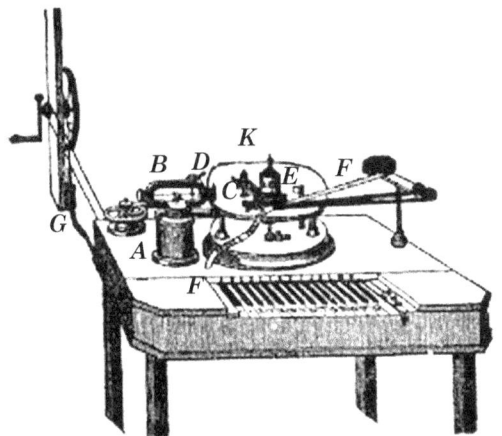

Fig. 5
House's printing telegraph

This is held back by a delicate spiral spring, graduated by a screw, which is also seen to the left. Above are two circular plates of glass. The upright bar, armed with two little knobs, to perform the part of a hammer, rises between these plates. When the armature is drawn to the magnet it strikes one of them, and, on being drawn back it strikes the other. As they are of different tone, the repetition of this signal at once draws attention to the register.

This call is similar in purpose or principle to those used by Sömmering in 1811, Schilling in. 1831, and Henry, Steinheil, and Wheatstone in 1836 and 1837.—(See *Davis's Description of Instruments*.)

In House's beautifully contrived printing telegraph—which prints messages in Roman letters—the electric current is merely employed for signalising, while the mechanical force of magnetism and atmospheric pressure is used to imprint the letters on a revolving piece of paper.

It is very difficult to convey a popular idea of its structure, or peculiar method of operation. We shall try to describe it so that its operation may be understood. A perspective view of the instrument is shown in Figure 5, comprising both the transmitting and receiving apparatus. The principle by which the different letters are signalised over the wire, is the transmission of a given number of electrical impulses for each letter, by the rapid opening and closing of the circuit. This is accomplished by means of the twenty-six letter-keys, and the two keys for the dot and dash, seen in the figure. Under the keyboard is a horizontal cylinder, which is kept in revolution by turning the crank and wheel, seen at the left of the figure. At one end of this cylinder is a circuit-wheel or break-piece, having fourteen projections and fourteen spaces, on which a spring, connected with the telegraphic circuit, bears. Consequently the battery circuit is completed fourteen times, and broken fourteen times, with each revolution of the cylinder. Under each key a projection or stop is placed upon the cylinder in such a position that when the key is depressed, and comes in contact with it, the cylinder shall have performed such part of a revolution as to have made and broken the circuit the number of times which represents the letter corresponding to the key. The motion of the cylinder is communicated by means of slight friction, and it is accordingly arrested by depressing the key. This is the transmitting or "composing" apparatus.

The receiving or printing apparatus is seen behind the keyboard in the figure. There is one such at each extremity of the line, to receive messages transmitted from the other extremity. But both are left constantly in the circuit, so that the operator signalises or prints the messages which he sends both at the distant end of the line and immediately before his eyes. The printing instrument, which we are examining, is, therefore, a facsimile of the one which receives the communication at a distance from the operator at the keyboard in the figure.

The printing apparatus consists of an upright rod-electromagnet, enclosed in the metallic cylinder A; of a little engine, operated by condensed air, and moving an escapement at B; of a type wheel at O; of a printing eccentric and lever, the end of which is seen at D; of a black colouring band at E; and the strip of printing paper at FF.

The electro-magnet consists of a compound rod of several short pieces of iron strung upon a rod of brass. This rod is enclosed in a tube of brass, attached to which, within, are several short tubes of iron, corresponding to and reacting with the pieces belonging to the axial magnet. This whole system of tubular and axial magnets is enclosed in a single helix of fine wire, connected with the telegraphic circuit. The tube is fixed, but the compound rod is movable, and attracted downwards by several co-operating reactions when the current passes. This rod is suspended by a cross wire, which may be seen stretched across the top of the cylinder A, and acts as a spring, drawing the rod back after the current has ceased to act. A very rapid vibration of the rod is thus obtained, corresponding to the opening and closing of the circuit effected at the transmitting end of the line.

Connected with the wheel is a condensing pump at G, which keeps up a supply of condensed air. At the upper part of the electro-magnetic rod is a collar-valve, which changes the direction of the current of condensed air with each vibration of the rod, though these vibrations are only one sixty-fourth of an inch. The air is thus admitted to opposite sides of the cylinder of a little atmospheric engine, which, by means of its reciprocating motion, permits the action of an escapement, tooth by tooth, and the corresponding revolution of the type-wheel, which is impelled by a spring kept wound up by the manual power employed at the crank and wheel.

The result is, that the type-wheel K, which has twenty-eight teeth, revolves just as far as the cylinder attached to the circuit-wheel, at the distant extremity of the line, has been permitted to revolve by depressing one of the keys. Each break, as well as each completion of the circuit, thus corresponds to a letter. It only requires that the instruments at both ends of the line should be set to the same letter, and then the cylinder at one extremity, and the type-wheel at the other, regulated by the pulsations of the current, will always revolve at the same rate; and if the cylinder is stopped at any one point representing a letter, the type-wheel is stopped at the same point, and presents the type which it carries on its periphery to the strip of paper in front of it.

When the type-wheel stops, an eccentric, actuated also by the local power at the crank and wheel, brings the black band and paper forcibly against the type, and leaves the impression of the letter. The paper is then carried on just the distance of a letter, and is ready for another impression. Roman letters are thus printed over a long line at the rate of from one hundred and fifty to more than two hundred a minute.

In the figure, the letter A will be observed at a little window above the type-wheel. This letter is on a letter-wheel, connected with the type-wheel below, so that the letters may be presented to the sight at the same time as printed; or the printing eccentric may be detached, and only the visible letters read.

The action of the electricity in this telegraph is merely to produce correspondence of motion in machinery at different ends of the line, in the same manner that uniformity of rate has been secured in clocks at different places, regulated by the electro-telegraphic current. All the mechanical results of House's telegraph are produced by local mechanical power. For this purpose, clockwork, having a regular rate, would be preferable to manual power.

Bain's Telegraph. — The decomposition of a metallic point in contact with chemically prepared paper, by the action of an electrical current, has already been referred to. The telegraph of Bain, represented in Figure 6, is constructed on this principle, and is the most simple now in use. The indication of the current takes place here without motion. The circular tablet, on which the writing is obtained, is moved by clockwork, at a uniform rate, under the wire, which constitutes the telegraphic pen. But the pen itself never stirs. It bears silently on the tablet, and as the eye observes the point of contact, now a blank space, and now a deep blue line, appears upon the retreating surface. This is the record of the intermitting current, sent over the wires from a distance.

In the figure, the clockwork which moves the tablet is seen on the right. Its motion is regulated by a flywheel above, the vanes of which can be inclined so as to present greater or less resistance to

Brief Explanation of Terms and of Instruments Illustrated

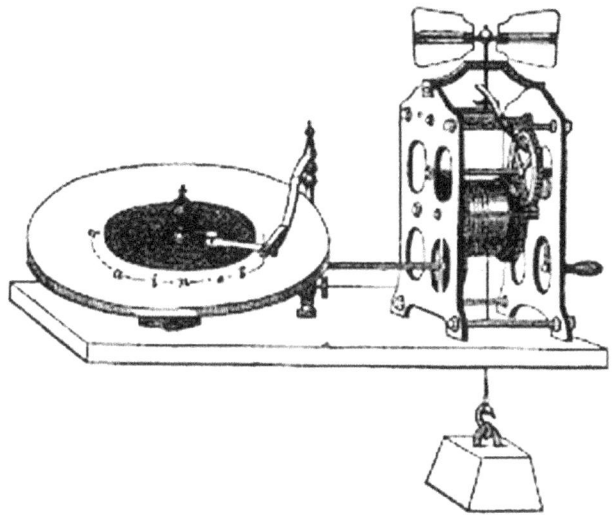

Fig. 6
Bain's telegraph

the air. A lever or brake bears upon the axle of the flywheel, by moving which lever the clockwork may be stopped, or allowed to go on. The circular disc, or tablet of brass, carried by the clockwork, is seen on the left of the figure, inclined towards the observer. In the centre of the disc, occupying the shaded portion, a spiral groove is cut, in which the guide to the pen travels. This guide is seen, attached at right angles to the penholder, which extends over the disc. The pen-wire is seen, held by a little clamp, descending so as to touch the tablet. This wire, of course, traces a spiral line of dots upon the outer ring of the disc's surface, exactly corresponding, in the distance of its lines, to the spiral groove within, which serves as a guide. By this beautiful contrivance, the writing is disposed in a close spiral line of dots, to represent letters, occupying but very little space.

The outer part of the surface of the disc, upon which the letters are represented in the figure, is covered with a ring of moistened and chemically prepared paper. This may be renewed or removed at pleasure. The penholder is connected with the positive wire of the telegraph, and the tablet with the negative. The circuit of conductors is completed by the moistened paper which intervenes, and which the current accordingly traverses. This paper is moistened with a solution of the yellow prussiate of potash, acidulated with nitric or sulphuric acid. The pen-wire consists of iron. When the current passes, this pen-wire is attacked by the solution, and the portion of iron dissolved unites with the prussiate of potash to form the colour known as Prussian blue, which permanently stains or dyes the paper.

A modification in the mode of marking has been introduced in this telegraph by Mr. Rogers, of Baltimore. He substitutes a pen carrying an ink which is decomposed by the current when in contact with the brass disc, without any intervening paper. A superficial stain is produced on the metallic surface, which is easily obliterated by friction.

In Bain's telegraph, no receiving magnet is necessary. The current traversing the long wires is sufficient to leave its trace upon the paper. The decomposition in Bain's instrument is instantaneous. This is an advantage over mechanical means to complete and break the circuit, as it secures greater speed for the purpose of rapid communication.

4
TELEGRAPH LITIGATION
EVIDENCE, DECISIONS, ETC.

In a subsequent part of this work will be found a brief sketch of the commencement, progress, and present length of telegraph lines in the United States.

Attempts to introduce instruments invented by House, Bain, and others, have been legally contested by Professor Morse and his associates, on the ground that they infringed his patent rights.

It would be impossible to give a rational account of the rise and progress of the electric telegraph in the United States without referring to these trials, had before the United States District Courts. Though a delicate task to give a clear analysis of cases in which so much feeling as well as pecuniary interest were involved, yet, in making the attempt, we have endeavoured to do so with impartiality. We have simply sought to arrive at the truth irrespective of party interest—sustained by the legal evidence of competent and respectable witnesses. That parties on either side will feel satisfied with the result, is more than we can predict. The litigation which has been so perseveringly followed up, in respect to electric telegraph rights has involved a large expenditure of money, both on account of plaintiffs and defendants. Some of the cases are still before the courts. Few patent cases have ever puzzled lawyers and judges more than those which have been prosecuted in this country, by Professor Morse and his friends, against other parties. Suits respecting ordinary patents are usually considered, both in this country and in England, to be surrounded with many difficulties.

In the case of electric telegraphs there are so many technical terms—the operations of the principal agent so occult—its development so ingenious—the mechanical and chemical contrivances so delicate as almost to elude the comprehension of the uninitiated. When the complexity of the subject is thus considered in connection with the legal technicalities of the patent laws, it will be imagined that the minds of judges must be strongly exercised to enable them to clearly understand the subject, and to arrive at a just decision.

The trials referred to have not been without useful results. A vast amount of valuable information has been elicited respecting discoveries in electrical and electro-magnetic science, and their bearing shown on the practical establishment of electric telegraphs. Without the evidence adduced in these cases, and afterwards printed, no such mass of information could have been elsewhere found compiled, so well calculated to elucidate the annals of the electric telegraph.

The first trial had, was on an application by Professor Morse for an injunction against Henry O'Reilly, to prevent his using an instrument invented by Zook and Barnes, of Cincinnati; and which was brought before Judge Monroe of the United States District Court of Kentucky, the result of which we have alluded to at another place.

The next action was commenced by F. O. J. Smith, chief proprietor of the New York and Boston Morse telegraph line, against Hugh Downing *et al*. Smith was said to be a partner in Morse's patent to the extent of one-fourth. Hugh Downing was the constructor and president of the line from New York to Boston, worked by House's printing telegraph.

Smith sued for an injunction against Downing before the late Hon. Levi Woodbury. The case was tried at Boston in 1850. The Hon. Judge Curtis, (Woodbury's successor) was counsel for plaintiffs, and Mr. Gifford, of New York, principal counsel for defendants.

The printed testimony taken in the case amounted to between 400 and 500 pages.

A large number of scientific witnesses were examined, among whom were Dr Wm. E. Channing, Professor Hare, Professor Joseph Henry, Secretary of the Smithsonian Institute, Professor Renwick, of Columbia College, Dr Chilton, Professor Silliman, Sen., Dr Chas. T. Jackson, besides many other men distinguished for chemical, mechanical or engineering skill. The leading weight of the whole testimony of the gentlemen given above, having any direct bearing, went to show that House's instrument did not infringe the patent claims of Morse.

The result was that Judge Woodbury, after due deliberation, denied the injunction, and decided that House's printing telegraph did not infringe the claims of Morse.

The greater portion of the evidence referred to above, which had any bearing on the subject, was reproduced at a subsequent trial had at Philadelphia in 1851, before Judge Kane, of the U. S. District Court for the eastern district of Pennsylvania, in the case of *Benjamin B. French et al.* vs. *Henry J. Rogers et al.* The plaintiff was president of the Morse New York and Washington line, and the defendant represented another New York and Washington line, worked by Bain's instruments. In this case, it was claimed, that the use of Bain's instrument was an infringement of Morse's patents; hence an injunction was sued for.

The evidence taken by complainants and respondents made about 1000 octavo pages, embracing all the leading and important evidence submitted on the previous trial.

Judge Kane, contrary to the expectation of almost everyone who had examined the subject, decided for plaintiffs, and granted the injunction. We shall show from the evidence submitted on the trial, in another chapter, the extraordinary character of the decision. The testimony adduced by the defence clearly proves it to be unsound and untenable.

Before, however, reaching the division of our work referred to, it may be as well to touch upon some of the facts stated by witnesses in the case of *Smith* vs. *Downing*, which are more interesting as illustrative of the history of telegraphs, than as important direct testimony.

Our chronological statement, in its main features, is sustained by Dr Channing in his evidence in the case of *Smith* vs. *Downing* at Boston. See p. 41 of the *printed testimony*.

In this case, also, Oliver Byrne, at p. 199, among other things gives the following statement:

> In the year 1830, I attended the public lectures of Abraham Booth (afterward scientific reporter for the Times newspaper,

and who became Dr Booth), delivered in Dublin, among other subjects, on electricity and electro-magnetism. In said lectures, the said Booth, in my presence, used in combination a long circuit of insulated wire conductors, or galvanic battery, an electro-magnet with an armature and mercury cups to join and disjoin the circuit, with which he magnetised and demagnetised the iron of the electro-magnet, causing it to attract the armature when the circuit was joined, and to recede from it when disjoined. Mr. Booth, at that time, stated to his audiences that that power could be produced and used at distant places, as signs of information; and he repeatedly illustrated what he meant, by causing the armature to approach the magnet, and then to fall from it on the floor, stating at the same time that it made marks by so falling.

Dr Channing, in respect to the order of discovery, has the following in his testimony referred to above:

Before the year 1750, the apparently instantaneous passage of electricity through several miles of wire, and even the use of the ground for the return circuit had been observed. Various telegraphs had been invented and described during the first period; and in 1798 Betancourt is stated to have operated successfully in transmitting signals on a line of 26 miles, between Madrid and Aranguez. (See *Sturgeon's Annals of Electricity*, vol. iii.) During the second period, telegraphs by free electricity were described and used, and also the galvanic telegraph of Sömmering and Dr Coxe, in which the indications were made for the most part by the decomposition of water.

Soon after Professor Oersted's discovery of electro-magnetism, Ampère described the first electro-magnetic telegraph in his memoir, in the *Annales de Chemie* of 1820; he describes the use of the deflection of the needle for signalising letters at any distance, by means of depressing signal keys to close and break the circuit. (See *Annales de Chemie et Physique*, vol. xv., p. 73: 1820.) The use of signal keys in combination with the telegraph was thus early invented and described.

At p. 40, S. vs. D., he states:

I believe the connection of the graphic register with the electric telegraph to have been made and published by Steinheil, in Germany, before the date of the caveat of S. B. Morse, October

1837. In the paper of Steinheil, included in the transactions of the *French Academy of Sciences* of the 10th September 1838, and published in the *Comptes Rendus*, of 1838, he described the results of the practical operation of his graphic telegraph, for more than a year previously, between Munich and Bogenhausen; and I understand the 19th of July 1837, to be referred to by him as an historical date, on or before which his electro-magnetic telegraph was in actual operation and public use.

In an article by Steinheil, translated in *Sturgeon's Annals of Electricity*, of March and April 1839, the use of posts for insulation, of what is technically called the ground circuit, and of iron instead of copper wires for conductors, facts or inventions of great importance to the practical operation of the telegraph, are fully described. I consider Steinheil, together with Gauss and Weber, who erected their telegraph at Gottingen in 1833 and '34, as the explorers of the electric telegraph, to whom the most important part of its practical application is undoubtedly due.

Professor S. F. B. Morse claims that the first time the idea of an electro-magnetic telegraph entered his mind, was as he came home from France in the ship *Sully*, in 1832. Dr Chas. T. Jackson, of Boston, being a passenger in the same vessel, claims to have imparted the first ideas of such an enterprise to Professor Morse, which the latter denies; and both have introduced elaborate statements and evidence to maintain their relative declarations, though, however, seeing that so much had been already done to establish the practicability of electric and electro-magnetic telegraphs prior to their crossing the Atlantic together, in 1832, we cannot perceive that the claims of either for suggestions at that time, possessed any originality, or had anything to do with the history of telegraphs.

Dr Jackson claims that he had a small electro-magnet with him, on board the packet-ship, and that he explained its character and functions to Morse, and suggested the practicability of an electro-magnetic telegraph to him. He also claims that he suggested to Morse some general idea of a mechanical contrivance similar to his port-rule and type, which Morse however, in his own examination denied, and introduced testimony in favour of his statement. Be this as it may, it will be seen by the evidence of Mr. Avery in another part of this work, that the type and port-rule, first contrived by

Morse, proved of no practical utility, and indeed proved a failure in attempts to reduce it to practice.

Professor Henry to whom so much is due for his discoveries in electro-magnetism, and who may be said to have been the first to discover the practicability of producing mechanical effects, or telegraph signs at great distances, gives much interesting testimony in the case of *Smith* vs. *Downing*. But as we have drawn largely from it in a future chapter, we will only in the present place refer to a statement made to Professor Henry, by Dr Levi Gale, who became associated with Professor Morse, and at one time was part owner of his patent, and who is now an Examiner in the Patent Office at Washington. Professor Henry says Dr Gale told him that, when he, Gale, first became connected with Professor Morse, about 1837, he, Morse, had not succeeded in producing effects at a distance; that when he first called in, he found Professor Morse could not by an electro-magnet, produce effects at the end of a copper wire, of a few yards in length, hung around a room in the University of New York. Dr Gale asked him if he had seen the paper published on the subject by Professor Henry, in *Silliman's Journal*, and he answered "No". He then informed Mr. Morse that he would find the principles of success explained in that paper—that instead of a battery of a single element, he should employ one of a number of pairs; that instead of a magnet with a short wire, he should use one with a long coil.

Mr. Gale stated that he had apparatus of the kind in the building, and that by applying it, action was produced through the wire for a distance of half a mile. — See page 92, *S.* vs. *D. et al.*

It may be well to notice the fact as a link in telegraph records, that Mr. Harrison G. Dyar in 1826 attempted to establish an electric telegraph at the race-course on Long Island, New York.

The account of Dyar's experiment is given by himself, in a letter to Luther J. Bell, Esq. living near Boston.

His letter to Mr. Bell is dated, "31 Rue de la Madeline, Paris, March 8, 1848". It is too long to give entire, which seems to have been written in reply to inquiries made by Mr. Bell, regarding his establishment of an electric telegraph, on Long Island, in 1826. He claims that Professor Morse's mode in representing the letters of the alphabet is similar to his own:-

> Since reading your letter, when searching for some papers in reference to my connection with this subject, I found a letter of introduction, dated the day before my departure from America, in February 1831, from an old and good friend, Charles Walker, to his brother-in-law, S. F. B. Morse, artist, at that time in Europe. At the sight of this letter, it occurred to me that this Mr. Morse might be the same person as Mr. Morse of the electric telegraph, which I found to be the case. The fact of the patentee of this telegraph being so identical with my own, being the brother-in-law and living with my friend and legal counsel, Charles Walker, at the time of and subsequent to my experiments on the wire, or electric telegraph, in 1826 and 1827—or about twenty years ago—has changed my opinion as to my remaining passive and allowing another to enjoy the honour of a discovery which, by priority, is clearly due to me, and which, presumptively, is only a continuation of my plans, without any material invention on the part of another. (See evidence in case of *S.* and *J.*, p. 20, A.)

Mr. Dyar, although deserving of much credit, did not accomplish any more, if as much, as others who had preceded him, and particularly Ampère, in France, in 1820. From his statement, it seems likely that Professor Morse may possibly, while in Europe, in 1831, have learned something of what Dyar had attempted, from himself, or of attempts made in Europe by leading scientific men, and other parties, or from scientific publications. And that Dr Jackson had also heard of the electric telegraph experiments of Ampère and others, and that both returning home in the same ship, having the subject in their heads, conversed freely about it; and that Dr Jackson, understanding more of electric science than Mr. Morse, who had been for some years devoted to the palette as a painter, the latter freely sought information of the former on the all engrossing subject. As far as the original discovery of the electric telegraph was concerned in 1832, neither party could make an exclusive claim to it.

Mr. Dyar says that he contemplated in 1826, extending telegraph wires on poles, through the air from New York to Philadelphia, but thinks he was ten years too soon.

> I invented a telegraph which should be independent of day,

or night, or weather: which should extend from town to town or city to city, without any intermediary agency, by means of an insulated wire in the air, suspended upon poles, and through which wire I intended to send strokes of electricity in such a manner, as that the diverse distances of time, separating the diverse sparks, should represent the different letters of the alphabet and stops between the words H and C. The absolute stops on the relative difference of time between the several sparks, I intended to take off from an electric machine by a little mechanical contrivance regulated by a pendulum, and the sparks were intended to be recorded upon a moving or revolving sheet of moistened litmus paper, which, by the formation of nitric acid, by the spark in the air, in its passage through the paper, would leave a red spot for each spark on this blue test paper.

These spots he proposed to be so spaced as to represent the letters of the alphabet, or for other signs to be transmitted, over any length of wire, backwards or forwards. He also proposed to employ an auxiliary aid along the wire to gain greater impulse—meaning probably, electro-magnetic power.

In reference to what I did to carry out my invention, I associated myself with Mr. Brown, of Providence, who gave me certain sums of money to become associated with me in the invention. We employed a Mr. Connel, of New York, to aid in getting capital wanted to carry the wires to Philadelphia; this we considered as accomplished; but, before beginning on the long wire, it was decided that we should try some miles of it on Long Island. Accordingly I obtained some fine card wire, intending to run it several times round the race course on Long Island. We put up the wire, i.e., Mr. Brown and myself, at different lengths, in curves and straight lines, by suspending it from stake to stake, and tree to tree, until we concluded that our experiment justified our undertaking to carry it from New York to Philadelphia. At this moment our agent, Mr. Connel, brought a suit or summons against me for $20,000, for agencies and services, which I found was done to extort a concession of a share of the whole project. I appeared before Judge Irving, who, on hearing my statements, dismissed the suit as groundless. A few days after this, Joseph E. White, who knew about our plan of a new telegraph by wire hung in the air, and who was our patent agent

Telegraph Litigation, Evidence, Decisions, etc.

(intending to take out a patent when we could no longer keep it a secret), came to Mr. Brown and myself, and stated that Mr. Connel had obtained a writ against us, under the charge of conspiracy, for carrying on secret communication from city to city, and advising us to leave New York until he could settle the affair for us, as the Sheriff's officers were then after us. As you may suppose, this happening just after the notorious bank conspiracy trials, we were frightened beyond measure, and the same night stepped off for Providence, where I remained for some time, and did not return to New York for many months, and then with much fear of a suit.

This seems to have put an end to Mr. D's experiments. On returning to New York, he thinks that he "consulted Charles Walker, who thought, however groundless such a charge might be, that it might give me infinite trouble."

In the late trial before Judge Kane, at Philadelphia, in the case of *Benjamin B. French, et al.*, representative of Morse's patent, *vs. Henry J. Rogers, et al.,* representative of Bain's patent, Mr. Dyar having returned to the United States, testified in person to the improvements he had made, pretty much in substance as stated in the extracts given from his Paris letter. He however declined to tell all he professed to know about it, or which he claimed to have discovered, on the ground that there were principles involved in it for which he was seeking patents in the United States and Great Britain. (See pp. 13-22 *French vs. Rogers*, September 1851.)

We shall next proceed to investigate the testimony given before Judge Kane, of Philadelphia, in September 1851. We shall show from the evidence, that there was not the slightest interference on the part of the Bain's Telegraph, with that of Morse. We shall show, that the great fault with Morse and his friends has been, to claim the exclusive use of principles which he never discovered or invented; and that all he could ever justly claim was his mechanical contrivances, to be operated upon by electricity, or electro-magnetism; such as his port-rule, and his pen-lever, in combination with electro-magnets and clock-wheels—his pen-lever to make dots and lines on slips of paper passing over a grooved roller, moved by the clock-wheels. There is nothing else used by him, that was not proved on the trial to have been discovered and used by others before he had ever taken out any patent at all. It is by the excessive and untenable extent of his claims, that his cause has been so much weakened.

There is no living man of an unprejudiced mind, who will take the labour of wading through the voluminous evidence referred to, who will not arrive at the conclusion that the decision of Judge Kane is one of the most extraordinary ever delivered in a court of justice. It was given in direct opposition to the direct testimony of more than half a dozen highly respectable scientific witnesses, bearing upon the direct points at issue; otherwise, how could he have decided that Steinheil's Telegraph was not a recording but a visual telegraph, when it actually recorded by dots on slips of paper, which was in evidence. His decision proves either one of two things: that his mind was probably made up without a careful examination of the testimony, or that he grossly erred in interpreting it. According to such a decision, no improvement, however great, important, or beneficial, can be attempted. A similar decision would have kept the steam-engine and printing-press where the first patent improvers left them. On this principle, if one man could patent steam, another air, another water, and another electro-magnetic motion—four powers, or elements—four patentees could become the exclusive owners of all the steam engines, wind-sails, windmills, water-wheels, and telegraphs, in Christendom.

A patentee has no right to use his discovery for the oppression of the people. A patent is given, not because he has discovered a mode of inflicting injury, but on the ground that he has conferred a benefit, and not a curse, as a telegraph monopoly granted on the broad decision of Judge Kane would prove to be. "If a man is honest, it is his duty in writing a history of the telegraph, to tell the truth, and to speak from facts adduced under oath."

We are personally friendly to Professor Morse, and esteem him as an amiable man, deserving of credit and a fortune, for his application and perseverance in the introduction of the telegraph. But were he our brother, in writing a history we should still feel bound to tell the honest convictions at which we have arrived—to utter the whole truth, and nothing but the truth. We have no private ends to subserve. We own no interest in any telegraph patent. We own not a dollar of stock in any telegraph line. Those for whose interest we write are the people; they have a vast deal at stake; and as one of the people, we have a right to examine the subject, and to urge our rights in the premises.

When a steamboat company, by a legislative act, claimed a monopoly for exclusively navigating all the rivers in the United States, it was thought to be oppressive. Professor Morse and his friends claim an exclusive right to navigate the air by electricity, over the whole continent. As broad as the claim is, we should be perfectly willing to accord it to them, did we not know, from incontestable evidence, that they were not entitled to it.

It may be stated, and can be indisputably proved, that, had neither Jackson, Morse, House, nor Bain ever existed, we should still have had, at this time, electric telegraphs in operation. The discoveries of Oersted and of Ampère, in 1820, of Henry, in 1827 and 1831, of Gauss and Webber, in 1833, with the erection of a working recording telegraph by Steinheil, in 1837, and of a magnetic telegraph, patented by Cooke and Wheatstone, in 1837 (all in advance of Morse's patent), placed the permanent introduction of electrical telegraphs beyond all doubt or dispute. Morse, and others who have followed them, have only added mechanical contrivances to what had already been invented or discovered, and for which alone they are entitled to patents.

We shall proceed, in our next chapter, to show from the evidence given in the trial of *French* vs. *Rogers*, all that had been known about electric telegraphs before Morse's time, in addition to what we have already adduced, and also show that there is no interference or infringement on Morse's patent by that of Bain's; and to prove what Morse is really entitled to, and what he has no claim to; with a synopsis of the evidence set aside or overlooked by Judge Kane in making his decision.

5
LEADING POINTS OF CONTEST
BETWEEN THE CLAIMS OF
MORSE AND BAIN

In the previous chapter we gave the substance of some of the evidence elicited in the trial of the case between *Smith* and *Downing*, or *Morse* vs. *House*. There was much new matter introduced at the trial between *French* and *Rogers*, or *Morse* vs. *Bain* in Philadelphia, in 1851. Mixed with the whole there was also a great mass of irrelevant verbiage embraced in the interrogatories and answers taken out of court, which without elucidating the case greatly increased the amount of printed testimony on both sides. To wade through a thousand pages of printed evidence is no light task. Although the printed testimony in the case of *Smith* vs. *Downing* only reached from 400 to 500 pages, yet Judge Woodbury took two or three months to weigh and examine it before giving his decision. Judge Kane of Philadelphia, on the contrary, in the case of *French* vs. *Rogers*, where the evidence on both sides made about 1000 printed pages, took the brief period of about four weeks to form an opinion and give his decision.

In this case the counsel for the plaintiffs maintained that Bain's patent infringed the claims of Morse in the following points:

1. They contended that it infringed Morse's claim to the use of dots and lines to represent the letters or signs of letters, at a distance by which communications were recorded.

2. That Bain's by using what he termed a branch current, infringed Morse's claim to what he claimed as a local circuit. Also, that Bain's by using connecting magnets, by which circuits might be opened and closed and messages sent to a greater distance, infringed Morse's claim. They also considered the use of the call

or signal, and receiving magnets as interferences.

3. That Bain's employment of chemically prepared paper to be marked by the passage of an electrical current in the decomposition of metallic wire in contact with the prepared paper, whereby black marks or dots were produced, was an infringement.

Before going further, it may be as well to state the chronological order of Professor Morse's patents and claims.

1. In October 1837, he entered his caveat in the United States, soon after which he went to Europe.

2. In 1838 he obtained a patent in France, but was refused one in England.

3. Having returned from abroad, he, in June 1840, obtained his first patent in the United States.

4. In January 1846, he obtained a reissue of his patent of 1840, in which he dropped his claim to the exclusive use of the Electric or Galvanic current, but claimed the use of Electro-Magnetism or Electro-Magnetic motion with other new matter.

5. In April 1846, he obtained a new patent, in which he claimed the exclusive use of the local circuit, which he employed for the purpose of increasing the mechanical force necessary to work his machine.

6. In 1848 he obtained a reissue of his patent of April 1846. In some of his patents he also claimed the use of the ground circuit.

7. In 1849 he obtained, under Commissioner Burke, a patent for a chemical telegraph, in which he claimed the use of paper chemically prepared with salts, to be acted upon by the passage of a current of electricity in causing their decomposition.

As Edward Davy, however, had taken out a patent in England, in 1839, for the use of chemically prepared paper, &c, for the same purpose (see *Respondent's Evidence*, p. 44), and as Bain's patent bore date in England, December 1846, (see *Evidence*, p. 157), Judge Cranch, at Washington, before whom the case was carried on an appeal from the Commissioner of Patents, decided that Bain was entitled to a patent; and it was issued to him accordingly. This point, therefore, did not appear to be insisted upon by

Morse's friends in the late trial, as an interference of very great importance.

From the chronological statement given, it will be perceived that in the same year that Morse entered his caveat at Washington, that Steinheil in Germany had actually put in operation a recording telegraph of several miles in length in July 1837, between Munich and Bogenhausen., and prior to the entrance of Morse's caveat in October following. (*See Defendant's Testimony*, p. 371.)

As stated, Professor Morse visited Europe between the entrance of his caveat in October 1837, and the entrance of his specification and obtainment of his patent in the United States, in 1840. We do not charge that, while he was abroad, that he availed himself of all that was then doing or known regarding electric telegraphs in Europe, which at that period were attracting great attention. There was not, however, anything to prevent his knowledge of all that Ampère and Steinheil, with Cooke and Wheatstone, had done, and were then doing. The latter parties took out their patent for a telegraph, in which they claimed the power of giving signals and sounding alarms at a distance, which was afterwards claimed in Morse's patent in 1840. Their patent was dated December 12, 1837. (See *Defendant's Evidence*, p. 33.)

On his return, therefore, to the United States, it was possible for him, if he chose to do so, to embrace in his specification in 1840, under new names, all the discoveries made by others, or himself, up to that period.

The propriety of allowing a patentee to obtain frequent reissues for patents, in which claims are to be expanded, has been seriously questioned. From the time Professor Morse entered his caveat in 1837, to the date of his last patent in 1849, it will be seen embraces a period of about twelve years. In this period he obtained four patents, and two reissues of patents. Before all the patents and claims of Morse expire concerning telegraphs, about fourteen years will have to pass over from the date of his last patent in 1849—making a total of about twenty-six years, from the entrance of his caveat in 1837, to the expiration of his last patent of 1849, in 1863!

If a patentee can continue to obtain reissues, and be permitted to expand his claims under each, it is clear that his operations might be made to cover half a century as well as a shorter period; and each intermediate discoverer could do nothing more than add to the interest of the one party!

It will be seen by the claims put forth by Morse and his friends in these various patents and reissues, that there was, apparently, an effort made to secure and to cover the whole ground of electric telegraphs in the United States to the exclusion of all competition, and for the sole benefit of himself and colleagues.

In England the electric telegraph has become a monstrous monopoly, being chiefly owned and worked by railway stock jobbers. The people at large are, in a measure, shut out from its benefits. Their monopoly was created by purchasing up patents from successive inventors, such as Davy, Bain, &c, and fighting weaker claimants in lawsuits.

In the United States it looks as though a similar monopoly had been attempted; not by the purchase of others' rights, but by the multiplication of patents and reissues made, to claim everything pretty much in the *lightning way*, and on these expanded claims to fight off all competition in constant lawsuits. In this, however, success has been only partially realised.

During the period of twelve years, from 1837 to 1849, the minds of many ingenious men in Europe and America were strongly attracted to the same subject, and the study of new improvements. These from time to time were made public; and it would be *possible* for a party under reissues, or new patents, to compile important facts from them, and to claim them under new names.

To this day, Cooke and Wheatstone's telegraph is substantially the only one employed in England. Morse, as stated, failed to obtain a patent in England; to which, however, we think he was entitled so far as the just merits of his mechanical combinations were concerned.

During legal proceedings had in various parts of the United States, the question has been mooted, whether Morse's first patent obtained in the United States, should not have borne date with his French patent of 1838. And in an action had before Judge McLean of Ohio, against Henry O'Reilly, this point was brought in question; and the Judge, it was said, was disposed to decide it in the affirmative, viz., that by dating up his patent two years from the date of his French patent to 1840, taken in the United States, weakened the validity of the American patent. Judge Woodbury, in his decision, alluded to it; but as it did not call for an opinion in the case before him, he passed it over.

The error with many witnesses for plaintiffs, whose testimony was taken in the case of *French* vs. *Rogers*, arose from the belief, that as electro-magnetic motion or force was employed to actuate machinery—a combination of which was contrived by Morse for telegraph purposes—no matter how generated or used, whether by one or more batteries, conveyed by one or more wires—or moved by one or more magnets of greater or less strength, or size; that, therefore, no other person, discoverer, inventor, or improver could employ electro-magnetic motion, however developed and arranged or employed in connection with different combinations, or with other mechanical means or modes for producing telegraph operations.

A large number and variety of chemical and mechanical contrivances have been introduced in Germany, and other parts of Europe, for electrical telegraph purposes—each differing in some respects from the others. Professor William F. Channing in his evidence for defence, page 373, says:

> The number of telegraphs belonging to the different periods had been so greatly multiplied at this time, or soon afterwards, that Quetelet states — on the authority of Wheatstone, in the *Bulletin de l'Academie Royal des Sciences de Bruxelles*, published at Brussels in 1838, p. 46 — that there were already at that time *sixty-two claimants* to the discovery of the electric telegraph.

In this country, as previously stated in our chronological table, besides the instruments of Messrs. Morse, House, and Bain, a Mr. Horn invented an *igniting telegraph*. That is, his machine was so arranged that instead of dotting, or marking paper, the spark of electricity actually burnt dots and lines through slips of paper while revolving through it, so as to stand as signs for letters.

We also saw the machine invented by a Mr. Johnson of the

western part of New York, for stamping dots on paper with leaden shot pressed against a slip of paper as it was drawn through it. Daniel Davis, of Boston, in his description of instruments, describes another telegraph which he terms an *axial telegraph*, actuated by magnetic motive power, something on the principle of Professor Page's plan of generating and applying it as a motive power.

Soon after Henry O'Reilly had built up his extensive lines of telegraphs in the West, connecting the western with the eastern cities, a party at Cincinnati, known as *Messrs. Zooks and Barnes*, invented an important modification of Morse's machine, and brought it into use in February 1846, at Louisville, Kentucky. They, we believe, were the first to introduce duplicate *fixed magnets*, between the poles of which, on either side, there played an electromagnetic needle placed in a copper coil. When the current passed through this needle it was deflected, or repelled, or attracted, as the operator desired, by the poles of the fixed magnets, and thus indicated signals, &c. Zooks and Barnes claimed that their fixed magnets absorbed the excess of atmospheric electricity, and enabled them to work during thunderstorms. In April following, Morse obtained a patent in which he claimed the exclusive "use of Electro-Magnetic Motion and Magnetism" of all kinds, however developed. (See his *Specifications*, p. 57, *Evidence for Defence*.) Morse soon after sued O'Reilly for an infringement, in using Zooks and Barnes' instrument. The case was argued before Judge Monroe, a strong personal friend of Mr. Amos Kendall, in the U. S. Circuit Court sitting at Frankfort, Kentucky, who decided against O'Reilly. An appeal was taken to the United States Court, where, we understand, the case is soon to be argued and finally decided.

To have a clearer understanding of the subject, let us consider what things (some of which have been claimed by Professor Morse) were discovered and in use before his first thoughts were directed to telegraphs, or prior to the year 1832. Among them were the following, briefly noticed in our chronological statement:

1. The Galvanic Battery, the germ of which was discovered by Galvani, a professor of Anatomy at Bologna, Italy, in 1790. Galvani's attention was called to the subject in a very curious manner. He

was preparing some frogs with which to make soup for his wife who was ill; and in doing so, discovered that every time he touched the leg of a frog with a knife, or scalpel, that it contracted or moved. This effect he made known, and it was soon after justly attributed to electricity. Zinc was soon after adopted (as in the pile of Volta) for the scalpel, and intervening slips of woollen cloth moistened in acids for the flesh or muscles of frogs, and electricity produced. From the discovery of Volta, have sprung all the improved galvanic batteries at present in use for telegraphic and other purposes. Hence the science of generating electricity by the decomposition of metals in acids, has ever since borne his name, or has been called Galvanism. Hence the terms "*Galvanic Battery,*" "*Galvanic Current,* "*Galvanic Electricity*" &c. Volta was a professor of Natural Philosophy at Pavia, in Italy. In 1800 he invented his pile, which was the first galvanic battery ever made, in which metals and acids formed a part of the arrangements.

2. Electro-magnetic motion was discovered by Professor Oersted of Copenhagen, in 1819, by passing a current of electricity at right-angles across the needle of a common compass, which deflected it.

3. The action of an electric current in deflecting an electro-magnetic needle, was discovered by Ampère in France, in 1821, and the same subject was experimented on by Arago, and Sir Humphrey Davy, about the same time. The latter's discoveries were coincident with those of Ampère. Mr. Faraday made further researches, and discovered the rationale and plan of forming magnets at will, by the action of an electric current. Indeed, it was about this time, it may be said, that the power of electro-magnetic motion was fully demonstrated and explained. And Ampère showed how it could be applied to telegraph purposes.

4. The magnets of quantity and intensity were discovered by Professor Joseph Henry of the United States, between the years 1827 and 1831. And their mode of construction, with their availability in connection with the use of batteries of intensity and quantity to produce signals or telegraph signs, or mechanical effects at a distance, were taught to his pupils. He also discovered the use of relay magnets, to open and close long circuits. He told his class,

that by the aid of his magnets he could cause a weight to strike a bell at a distance, and thus make it answer the purpose of a telegraph.

With all this knowledge on the subject which had been developed and published to the world before Morse's time, or prior to 1832, what was there left for Morse to invent to make a telegraph?

Professor Henry, in his examination in the case of Smith *vs.* Downing (see *Evidence*, p. 93), in answer to this very question, answers as follows:

> Merely to combine his port rules, or pin lever, with a spring to be moved by the armature of a magnet, and to operate on slips of paper passing between rollers.

But, it may be asked, does this exclude all other parties from taking these same electrical and magnetic principles and contrivances, discovered in the order we have named them, and prior to 1832, the year Morse's mind was first directed to the subject, and applying them to produce a telegraph by different mechanical combinations or chemical arrangements? We think no candid man can fail to answer in the negative.

A still more extraordinary feature of Judge Kane's argument was, that Morse was entitled to protection because he had discovered "*an art*"! —a thing that Morse himself had never claimed. In all his specifications, and through all his patents and reissues, he only claims to have invented a "new and useful improvement."

To have invented telegraphing as *an art*, he should have discovered all we have named, or be the author of all *Galvani, Volta, Oersted, Arago, Ampère, the Davys, Faraday, Henry,* and *Steinheil,* with a host of others, had done before him, embracing nearly half a century. He should have invented the galvanic battery, discovered electromagnetic motion, discovered the plan of producing electromagnets, and of varying their power, and adapted them to making signals at a distance. Indeed, he should have discovered nearly all known in electricity, and been the first to suggest its application to telegraphing. These things neither he himself, nor his most sanguine friends, ever wholly claimed for him. Then, he has invented no art. Indeed, Professor Henry says (see *Evidence,* p. 90, *Smith* vs. *Downing*):

> I am not aware that Mr. Morse has ever made a single original

Leading Points of Contest between Claims of Morse and Bain

discovery in electricity, magnetism, or electro-magnetism, applicable to the invention of the telegraph. I have always considered his merit to consist in combining and applying the discoveries of others, in the invention of a particular instrument and process for telegraph purposes.

To settle this point, suppose Prof. Morse had never existed, would we now have in operation an electric telegraph? Certainly we should. Wheatstone's telegraph could today be introduced into the United States, without in the slightest degree interfering with Morse's patents.

In the trial of *French* vs. *Rogers*, a vast amount of time was consumed, and much evidence taken, to prove Morse's claims to what he actually invented, in the way of his mechanical combinations, &c., and what nobody disputed; and, also, to prove what was equally useless, that the principles of electro-magnetism and electricity, previously discovered, were necessary to actuate his mechanism.

We shall next examine the grounds of the defence, set up in the recent trial of *French* vs. *Rogers* (or *Morse* vs. *Bain*) and mainly in reference to the points claimed as infringements of the former's rights.

1. *As to the dots and lines.* The defence showed that Mr. James Swaim, of Philadelphia, in 1829 invented the use of dots and lines, which he termed a Mural Telegraph; the design of which was that parties separated might communicate with each other, by making lines and dots on the wall, by any sounding or pointed instrument, intended, by the number of dots or blows, and straight lines or scratches, to represent the letters of the alphabet and numerals. He published his plan with an alphabet in 1829, and it does not differ materially from that of Morse's, except that Swaim made his straight lines perpendicularly, instead of horizontally as made by Morse. (See *Defendants Evidence*, pp. 113, 114.)

Professor Steinheil, of Munich, is said to have been the first who actually put in operation a recording electromagnetic telegraph. This was in evidence for the defence, and testified to by a number of witnesses.

Professor Augustus A. Hayes, a practical chemist of Boston,

states, pp. 332 and 333, as follows:

> Steinheil used a single circuit of insulated wire on posts terminating in a helix, which was polarised by a magneto-electrical machine, and caused deflection in two magnetic bars, each of which bore a cup of ink opening through a capillary tube against a moving fillet of paper; *thus recording his telegraph signs.*

Again he says:

> As dots, spaces and lines, are more easily made by a simple application of the two motive forces of the clockwork and electric current, magnetising and demagnetising the soft iron (shaped like the letter U), so their use was most obviously suggested. But dots and spaces for cipher-writing have been in use, with arbitrary significations, for many years. In *Rees Cyclopaedia*, vol. viii., article *Cypher*, is given an account of several; and, among others, one invented by William Blair, in the year 1807, in which dots and spaces are so contrived as to represent an alphabet of 81 signs, each sign being interpreted by letters of the Roman alphabet. Steinheil, too, used dots and spaces to signify letters as described in his patent above referred to; and it may be suggested, that the specimen of telegraphing writing exhibited by Morse (as by Vail's Book) in 1837, might have been actually done by Steinheil's machinery.

Again:

> I regard the invention of recording machinery as only an invention of the particular means, not the discovery of a new *art.*

He further says:

> But, again, I do not consider that Morse was the first to use the influence of electricity, either as a motive power, or as a chemical agency to imprint signs of intelligence.

He then states what had been done on this subject by Dyar, Davy and others, before Morse.

At p. 18 of *defendant's evidence* Harrison Gray Dyar says, in allusion to his telegraph:

> "These signs were indicated or recorded on paper, by the action of the electric spark at the further end of the wire, passing

through the paper before the sparks were conducted off the wire into the ground."

Professor W. B. Benedict, Professor of Mathematics in the U. S. navy, and formerly a Teacher of Chemistry and Natural Philosophy, at p. 246, *defendant's evidence*, testifies to Steinheil's recording telegraph, and says that it could not only make dots, but that his process was suitable for recording the continuance of dots into lines.

Also see evidence of Professor William E. Channing, a cultivator of practical science, at p. 373, as to Gauss and Weber's discoveries; and that the former was the first who showed the practicability of using signs; and that in 1833 they actually "constructed the first simplified galvano-magnetic telegraph." See also his evidence at pp. 373 and 375 for defence. At the latter page he shows that similar signs, or dots and lines, were used in an anemometer, or register of winds, which was invented by one D'Ons en Bray, and published by the *Royal Academy of Sciences* at Paris, in 1734. This instrument used clockwork to carry fillets of paper over a cylinder with metallic points. A second cylinder, revolving near the first, dotted the paper whenever it revolved against the metallic points with the paper between; —the number being 32, to correspond with the points of the compass. See his testimony on p. 276, where he describes other similar registering or dotting machines, one of which was a wind, and another a tide register. See also pp. 378 and 379, same evidence. See *Evidence of Professor O. W. Gibbs*, Professor of Chemistry in the New York Free Academy, who at p. 437 sustains the evidence of Professor Channing as to the discovery of the electro-magnetic telegraph in 1833, by Gauss and Weber, at Gottingen; and that they were the first to conduct the wires through the air over house-tops, &c.; and that the proposal to use the ground for a circuit was first made and executed by Steinheil; and that he was the first to put in actual operation a recording telegraph, by *making* dots, &c., on fillets of paper. The first employment of a recording telegraph was also claimed by Amyot, in the *Compte Rendus* of the French Academy for 1838, July 9th, vol. vii., p. 82. (See *B. A. Gould's Evidence for Defence*, p. 68.

We here annex the signals arranged by *Swaim, Steinheil, Morse, Davy,* and *Bain*.

The Electric Telegraph in the United States

No. 1.—*Example from Swaim's in 1829.*		No. 2.—*Example from Morse's in 1840.*		No. 3.—*Example from Davy's in 1839, and Bain's in 1846.*
–	a	–	e	A . —
– –	b	– –	o	B . — .
– – –	c	– – –	a	C . . .
– – – –	d	– – – –	h	D . . —
|	e	—	l	E .
| |	f	— —	x	F — — . .
– |	i	– —	i,y	G . — . .
– – |	o	– – —	w	H . — —
| –	end of number.	— – –	m	I . .
| – –	end of question.	— –	n	J . . . —
– | –	I understand.	– — –	s, z	K . — — .
| – |	again.	— – —	k	L . . . .
| | –	end of spelled word.	— — –	t	M . — . —
				N . . — —
				O —
				P — . —
				Q — . — .
				R — — .
				S — . .
				T — . . .
				U — .
				V — — — .
				W — — . .
				X — . . —
				Y — —
				Z — — — —
				& — — — — —

Fig. 7
Signals as arranged by Swaim, Steinheil, Morse, Davy, and Bain.

Dyar claims having used dots in 1826.

- I add some letters from Steinheil's arranged signs, to better show the close agreement, in the primitive idea arising to the mind of each person using such means.
- Swaim addressed the senses of the recipient by scratches on a hard substance for the lines, and knocks for the dots, doubtless

leaving visible signs, as happens in Morse's operations. The book of Swaim purports to have been printed by Clark & Raser, Philadelphia, 1829." — (See *Evidence of Dr Augustus A. Hays for Defence*, p. 337.)

A	*B*	*CH*	*D*	*E*	*F*	*G*	*H*	*M*	*N*	*I*	*O*	*R*	*S*	*T*
•	•		•		• •	• •	• • • •	• • •	• •	•			• •	•
• •	• •	• • • •	• •	•		•						• • •	• • •	•

Fig. 8
Letters from Steinheil's arranged signs

- "The alphabet used by Bain is the same in principle as that employed by Dyar, Steinheil, and also by Morse, consisting of combinations of dots and lines." — (See *Davis's Telegraph Book of Instruments*, P- 29.)
- Steinheil in 1837 used dots and spaces as signs; their signification being dependent entirely on their position, the duration of the electrical influence, its cessation, and the number of the changes. I consider these different resorts as founded entirely on the adoption of existing knowledge of means, and not the production of ideas in even a new form.
- When Steinheil described his telegraphic combination, he made no claim to any exclusive property in the variation he made in existing arrangements of signs; and his modification carries with it internal evidence that such resorts are mere expedients, conventional with patentee and operator.
- "Between Swaim's and Morse's we see very slight differences only arising from one marking longitudinally and the other laterally. In Steinheil's, with instances of identity with Morse's, we also notice a far more systematic arrangement in the varied positions of the dots. If the ordinary machines for unrolling a fillet of paper are used, no dots appear; as even momentary contact of the point or pen, leaves a shorter or longer line.
- "The claim to these signs in combination with machinery is therefore deemed without foundation in essentials, and of a character unknown to the laws of patent inventions. The machinery as a combination of two parts may be new, so far as has been before stated; but as a whole, it does not produce any new result,

by new means of a mechanical or electro-galvanic character." — (See *Evidence of Dr Augustus A. Hays for Defence*, p. 337.)

Having shown from the evidence of so many respectable witnesses that Bain, in using dots and lines differently arranged and developed from those used by Morse—the former being the result of chemical instead of mechanical action—did not interfere with Morse's claims, we shall proceed to the next division of our subject, viz.: the claims of Morse and friends to the use of local circuits, relay and receiving magnets, &c.

6
LOCAL CIRCUITS
RELAY AND RECEIVING MAGNETS

The second points at issue, as previously stated, were the exclusive claims of Morse to the use of local circuits, and relay and receiving magnets; the local circuit generated from a small separate battery being intended by him to increase the mechanical action of his machine in making dots and lines on slips of paper, drawn by clockwork between small rollers.

Here again the evidence is more voluminous and conclusive against his exclusive claims for electro-magnetic motion, local circuits, and relay and receiving magnets, than it was found to be with regard to the dots and lines.

Professor B. A. Gould of Boston, p. 68, for defence says:

> The earliest proposition of a local (that is, of a circuit for giving signals by the action of another) circuit that I have seen, is in the specification of Cooke and Wheatstone, for which an English patent was issued June 12, 1837.

Again, Professor Renwick of New York, in his evidence for plaintiffs, p. 492, says:

> I have examined and believe I understand the electric telegraph of Edward Davy, described in the *Repertory of Patent Inventions*, new series No 67, London, July 1839. It does contain a continuation of circuits whereby one opens and closes another.

On referring to the specification of Edward Davy, at p. 44 *Evidence for Defence*, enrolled January 4, 1839, we find that he claims as follows:

> Fourthly, I claim the mode of making telegraph signals, or communications from one distant place to another, by the employment of relays of metallic circuits, brought into operation by electric currents.
>
> Fifthly, the adapting and arranging of metallic circuits in

making telegraph communications or signals by electric currents in such manner that the person making the communication shall by electric currents and audible apparatus, regulate or determine the place to which the signals or communications shall be conveyed.

We thus have here, announced in 1839, the claim to the use of local circuits, or branch or relay circuits, and patents issued for the same, prior to Morse's patents obtained in the United States in 1840 and 1846, in the latter of which he sets up an exclusive claim for the use of local or co-operative circuits and sues Bain for an infringement, in 1850; that is, some six or seven years after Davy's patent, we find Commissioner Burke at Washington giving Morse a patent for the same thing as an original discovery. But Davy was not the only party who claimed the first use of co-operative circuits and magnets. The same was claimed in the patent of Cooke and Wheatstone, dated December the 12th, 1839.

Professor B. A. Gould for the defence, p. 83, says:

> A magnet such as Morse uses for a receiving magnet is no more nor less than an electromagnet as already defined. Cooke and Wheatstone invented the receiving magnet which operates by deflection.

Again, same page, he says:

> Cooke and Wheatstone's receiving magnet has devices for performing analogous functions (to Morse's), so far as their telegraph requires it.

Professor Thomas D. Rogers, professor of chemistry in the medical department of the University of Pennsylvania, says in evidence for defence, page 89:

> The local circuit (of Morse) operates electro-magnetically; the branch circuit (Bain's), electro-chemically. The local circuit's effect when the circuit is closed, is entirely independent of the main line; whereas, the branch current is a part of, affects, and is affected by the main line.

Professor Sears C. Walker, attached to the U. S. coast survey, says page 199, in his evidence for the defence:

> No principles used in the telegraph are original with Professor Morse; only mechanical contrivances.

Local Circuits—Relay and Receiving Magnets

Again at p. 201, he says:

> I have examined the branch circuit used by defendants (Bain's), used in their line between Washington and New York. It differs in using a different law of Nature, namely, that which regulates the chemical action of the galvanic current, in producing discoloration of paper by decomposition of salt, or by a discoloration of a fluid on a metallic surface. It differs in form and structure, in being one of two duplicate or branch circuits passing through the same battery; having one portion of the circuits in common, which by being broken would break the circuit of both; whereas the main and local circuits in Morse's patent have each an independent battery, and no portion of the circuits in common, being broken would break the circuit of both. Another difference is that neither of Morse's independent circuits can possibly modify the quantity of the other; whereas, each branch in Rogers's (*defendants*' or Bain's) branch circuit modifies the quantity of the other branch.

Professor George Mathiot, employed in the United States coast survey as an electro-metallurgist, testifies in his evidence for defence, p. 208:

> The parts of the telegraph were invented and discovered to the best of my knowledge and belief as follows. To Grove belongs the battery (that is, the latest and best constant battery). To Steinheil we are indebted for first using a single conductor with many stations, and showing the mode of using it, and also using the ground for a return circuit, and also, to him I conceive belongs the Register and Stenographic Alphabet. To Henry is to be attributed the magnets and local circuits, batteries, and receiving magnets.

On p. III, he also confirms the fact, that Davy in 1839 used a local circuit, and recorded by "the joint action of electromagnetic and chemical power."

At p. 219 he gives it as his opinion, that Wheatstone's plan of using receiving magnets, "must eventually be used in the Morse offices when they come to work on very long lines without relays. I have tested the superiority of this form (the **U** form) of receiving magnet." See also pp. 223 and 227. The function of the branch circuit in Bain's plan, is rapidly to oxidise the iron point in contact with chemically prepared paper, and thus produce dark spots, or dots. The function of the local circuit employed by Morse is to increase the mechanical action of his machine.

Professor R. Keith, Professor of Mathematics in the United States Navy, sustains the evidence of all the witnesses hitherto given. To Steinheil he gives the credit of forming a registering telegraph, and to Cooke and Wheatstone the credit of first introducing co-operative circuits and receiving magnets (see pp. 232, 233); and that Davy, also, had used local circuits. At p. 234, *evidence for defence*, he says:

> *I have examined Cook and Wheatstone's patent. It contains in substance and principle the local circuit used by Morse.*

On the same page he says:

> *I have examined and do understand the electro-telegraph* of Edward Davy. It contains in substance and principle the same local circuit."

He also on pp. 234, 235, states that the branch circuit used by Bain, differs entirely in its plan and arrangements from that of Morse, and that while the latter is distinct from the main line, and is merely intended to co-operate mechanically, the former regulates the action of the main line, and is used simply to produce chemical decomposition.

Professor Wm. B. Benedict, at pp. 241, 243, sustains the previous witnesses for defence in every respect. Speaking of the branch circuit used by Bain, and the local circuit used by Morse, he deposes that:

> They are contrivances as essentially different in principle, form, and mode of operation, as the common principles which they both involve will allow.

Professor Joseph Henry, Secretary of the Smithsonian Institute, sustains the evidence given by the preceding witnesses. He says for defence, p. 264:

> The object of Professor Wheatstone, as I recollect it, in bringing into action a second circuit, was to provide a remedy for the diminution of force in a long circuit.

Professor Henry's electro-magnetic investigations commenced

Local Circuits—Relay and Receiving Magnets

in 1828, when he repeated Oersted's and Ampère's experiments. He succeeded by an improved arrangement in constructing a more powerful magnet than had hitherto been made, which he explained to his class in 1829. See p. 251 as above, also 252. He soon after discovered the plan of making magnets of intensity, by using a single long copper wire coiled round a piece of iron in the horseshoe form, connected with a battery of a number of pairs of plates, and a quantity battery by wrapping a similar piece of cold iron with several coils of shorter pieces of copper wire, used in connection with a single pair of plates.

> The first of these magnets, says he (p. 252), which is the one now employed in the long or main circuit of the telegraph, may be called an intensity magnet; and the second, which may be used in the local circuit, may be denominated the quantity magnet.

These discoveries of Professor Henry were published in *Silliman's Journal of Science and Arts*, in 1831, and which it seems, from testimony previously adduced, were not known to Morse until brought to his notice by Dr Gale in 1837, and without the use of which he could never have succeeded in producing signals at any considerable distance by his plan of mechanism. The local small battery, and duplicate wire magnet, discovered by Professor Henry, was patented by Morse in 1846, as a sort of exclusive property in connection with telegraphs. Both the kinds of magnets described by Henry are at this day employed in the Morse lines. In the early attempts to produce signals at a distance, it was found that in the transit of the electric fluid it had a tendency to waste or suffer so much diminution as to become useless.

Professor Henry states (p. 252) that, in experimenting, he found a great diminution of the fluid to ensue from the increase of distance, when using a battery of a single pair (or quantity); but by removing it, and substituting one of intensity, consisting of 25 pairs, he found that he could cause the current to act upon a magnet at 1000 feet, without any apparent diminution of power. "This," says he, "was the first establishment of the fact, that a galvanic current could be made to develop magnetism in an electro-magnet at a distance, and produce mechanical effects; and also of the means by which such transmission could be accomplished. I saw that the electro-magnetic telegraph was now practicable; and in publishing my experiments, including those just mentioned, I stated that the fact above mentioned was applicable to Barlow's project of such a telegraph," p. 252. Barlow's attempt with a battery of a single pair was made at Greenwich, England, in 1825; and his failure was caused by the diminution of the electrical fluid in its transmission, while Professor Henry succeeded, by taking one for the main circuit of intensity. Henry, in another part of his testimony, also describes Wheatstone's plan of opening and closing his circuits, by the deflection of a needle, which he witnessed during his visit to Europe in 1837.

Page 254, as above, he says that, after returning from Europe, he repeated the experiments of Daniel, of London, in forming a constant battery, which he had done in 1836, and of Steinheil's plan of a telegraph devised in 1837; and that he formed a telegraph of several hundred yards in length, which worked successfully through the college grounds at Princeton.

> I find, by reference to my notebook, that the date of these experiments was October, 1842, previous to the unsuccessful attempt of Morse to transmit currents of electricity through wires buried in the earth, between Washington and Baltimore, and before he attempted to use the earth as a part of the circuit. Previous to this time, and after the above-mentioned experiments, Mr. Morse visited me at Princeton, to consult me on the arrangement of the conductors. During this visit we conversed freely on the subject of insulation and the conduction of wires. I urged him to put his wires on poles, and stated to him my experiments and their results. (See p. 254, Evidence for Defence.)

Again on p. 255 he says:

> I heard nothing of the secondary circuit as a part of Mr. Morse's plan, until after his return from Europe, whither he went in 1838. It was not until after this Mr. Morse used the earth as a part of the circuit, in accordance with the discovery of Steinheil.

Professor Henry in 1831 and '33 explained how that, by raising

Local Circuits—Relay and Receiving Magnets

a magnet and letting it fall at the distant end of a wire, he could make it strike a bell, and thus produce signals or alarms. Mr. Morse claims the same thing in his specification for a patent in 1840, and reissue of 1846. Morse twice failed in his attempts to establish an electro-magnetic telegraph through the agency of his mechanism, and only succeeded finally by availing himself of what others had discovered.

He failed in the University of New York, in 1837, to produce, by the ordinary battery and magnet, mechanical effects at a distance. So says Dr Gale, as previously stated. He then succeeded by the use of what Henry had discovered. He again failed in 1844, by attempting to convey wires isolated in a leaden tube underground, between Washington and Baltimore. He then afterwards succeeded by insulating a single wire on poles, and using the ground for the return circuit, which Gauss and Weber, and Steinheil, with Cooke and Wheatstone, had actually accomplished in Europe before him. And had Congress never given the $30,000, and Morse made no attempts, we should still have had telegraphs at work in the United States. Their success had been placed beyond contingency by others in advance of Morse. Page 260 (see *evidence for defence*), Professor Henry states, when speaking of the Morse's local circuit compared with Bain's branch circuit, that the latter "is applied to produce a record by means of electro-chemical action; and, in that of Mr. Morse, to give motion to mechanical marking apparatus. In the case of the Morse arrangement, the increased power is produced by calling into action a separate battery, to give motion, by the means of the electromagnet, to the mechanical apparatus. The arrangement of defendant's (Bain's), in some respects, appears to me to be a different contrivance, more simple, convenient, and less expensive than the arrangement of Mr. Morse." See his testimony also at p. 263, on his cross-examination, where he explains the plan of using the local circuit which he had devised, by which, in combination with a quantity magnet, he could, by renewing and breaking the circuit, raise from and let fall weights to the floor. This part of his combination he repeatedly explained to his class from 1833 to 1848.

I also accompanied the exhibition with a statement, that

the same effect could be produced by the action of a battery at a distance, by ringing bells, or producing other mechanical effects. The results of my first experiments, in causing an electro-magnet to act through a long wire, furnished me with the means of accomplishing this. For this purpose, it was only necessary to attach a forked wire to the armature of a small intensity magnet, connected with the long circuit, in which was also an intensity battery. When the current was passed through the long wire, the armature would be attracted upwards, the short circuit would be broken, and the weight fall. I do not recollect to have exhibited the last part of this arrangement in my lectures, or remember when I invented it; but the invention was made and explained to others before the publication of Mr. Morse relative to his telegraph. The object of this invention was to illustrate the production of mechanical effects at a distance, by means of a long circuit being made to open a short circuit. From my previous experiments in the transmission of electricity through long wires, I was well aware of the fact that I could not cause the large quantity magnet to act by a battery at a distance, directly through a long wire; and hence the necessity of this invention to produce the effect which I always said I could produce.

Professor Jackson, of Boston, in his evidence for defence, p. 315, sustains the testimony of the previous witnesses. He alludes to the local circuit claimed by Morse in his patent of 1846—and for which he obtained a reissue in 1848. After describing it, he gives the sole credit of its discovery and use for telegraph purposes to Cooke and Wheatstone, and Edward Davy, from six to nine years previous to Morse's patent of 1846. He gives to Professor Henry the credit of having discovered the principle and its application of making one circuit open and close another, by arrangements previously noticed. Again, on the same page, he says:

> Edward Davy describes a short circuit which was used to record his signs, in consequence of the exhausted state of electric energy in the long circuit. This is likewise in principle precisely the same as Morse's. The self-stopping apparatus, mentioned in the patent of Morse for 1846, seems likewise to be described in the said patent of Cook and Wheatstone, and the patent of Davy.

Local Circuits—Relay and Receiving Magnets

Professor Hayes, practical chemist and assayer to the State of Massachusetts, in his evidence for defence, pp. 332, 333, 334, 335, 339, 341, and 343, sustains the evidence of all the previous witnesses. In alluding to the claim of Morse for the local circuit, in his patent for 1846 and its reissue in 1848, he says, p. 335:

> The combination of circuits (claimed by Morse), it is hardly necessary to state, had been earlier adopted, and is in strict accordance with expedients resorted to in general machinery, as well as in applying electro-magnetic influence. In the patents of Cooke and Wheatstone, and Davy, similar combinations are described; and the inference is, that the helices of these inventors are really of more practical value than the combination referred to in the Morse patent.

At p. 339, the Professor states, that Morse's claim, set forth for an apparatus which he calls a receiving magnet, "is new only in name—it being the reproduction of the coil, or helix, by which the attractive force by position is rendered more efficient, and was in general use for this purpose before 1838." He further declares in effect, that all the claim or title Morse has to the apparatus called receiving magnets, local circuits, and register magnet, is to the new names he has given to them—and under which he has claimed them in his patents, although discovered and used by others before him. Again, he contends that Cooke and Wheatstone's patent of June 1837, contains a description of an apparatus which embodies all that is new or useful in the combination claimed by Morse. Again, we find E. Davy in 1838 had described an arrangement for the same, published in 12th vol. *Repertory of Patent Inventions*.

Dr William F. Channing, p. 276 for defence, states, that the local circuit described by Morse in his patent of 1846, was first made and invented by Prof. Joseph Henry, at Princeton College. He also sustains all the other witnesses in regard to dots and lines, co-operative circuits, receiving magnets, and register magnets, &c., as having been discovered before Morse by Henry, Steinheil, Cooke and Wheatstone, and Davy, &c. He also states that Davy had provided for a complete system of relays of long circuits. "Cooke and Wheatstone provided for the multiplication of indicating instruments and of receiving instruments, with local circuits at way-stations in any part of the long or telegraph circuit where it might be desirable to repeat or multiply copies of a communication." (See p. 377.) "In the patent of S. F. B. Morse of April 11th, 1846, as reissued June 13th, 1848, I find a description of and claim to a receiving instrument applied to operate a local circuit, which had been patented in principle by Davy in England, July 4th, 1838, as above described." He states further, on the same page, that the local, or office circuit, may apply to some systems of telegraphs; but "that the decomposing telegraph invented by Bain forms an exception."

Morse, in his deposition in the case against House's patent, states that in communicating a description of his telegraph to the French Academy in September 1838, "it did not include the office circuit or receiving magnet, the utility of which was then unknown." (See p. 377.)

Dr Channing states, on same page, that when this declaration of Morse was made on the 10th September 1838—Cooke and Wheatstone's patent fifteen months before, and Davy's patent more than two months before, included the office or local circuit, the receiving magnet as well as the principle of relay. " Morse in his description referred to, gave no description of a receiving instrument, or either a local circuit or system of relays."

Thomas C. Avery, a philosophical instrument maker of New York and, at one time attached to the Philosophical Department at West Point, at pp. 432, 433, 434, and 435, in favour of defence, deposes to having been an assistant of Prof. Morse in his first attempt to lay his telegraph in leaden tubes underground, between Baltimore and Washington. Morse, he says, coated four wires and inserted them in leaden tubes, and buried them for a distance of ten miles. Avery previously urged his disbelief in the success of the plan, while Morse advocated it; and said he intended to employ one wire for transmitting messages, and the other for returning an answer. This was in December 1843, four or five years after Cooke and Wheatstone and Steinheil had put telegraphs in operation in Europe, the latter using a single wire, and the ground for a return circuit. The laying of ten miles of tubes cost about $10,000 of the $30,000 which Congress had appropriated, and which Mr. Avery

Local Circuits—Relay and Receiving Magnets

says, on trial proved a dead failure.

Page 434, he says:

> Upon making our experiments on the line, almost everything was a complete failure. The original plan of laying the wires in pipes underground was a total failure, and was abandoned. The battery with siphon tubes was a failure and abandoned. The relay magnets, six in number, were not sufficiently delicate. I was obliged to alter the journals to work on agate points, which made them work with more delicacy, and thereby Mr. Morse was enabled to work with a half quantity battery. The adjustment of the agate points was my own invention. Mr. Morse was not present, Mr. Vail was, but did not know what an agate point meant. The instrument made for writing never was used, being found totally useless; it was cut up by me, and part of the materials used by me for making the new improvements. The recording instrument had also to be altered.

Mr. Avery states, p. 433, that, during the experiments with the lead pipe, he told Mr. Morse that in Europe they put the wires on poles. The leaden tube plan was then abandoned, and, he continues,

> About the first of April 1844, we commenced to put up the wire on poles. We erected two wires from Washington to Bladensburg, a distance of eight miles. The two wires could be connected, so as to make one circuit. An experiment was then tried on one of the wires with plates in the ground as part of the circuit. I then telegraphed a message to Washington.

This was the first telegraph line actually put in operation in the United States. Had Congress offered a premium for the successful erection of a telegraph line of 40 miles, between 1838 and 1844, one would have been in successful operation in advance of Mr. Morse.

Mr. Avery states, that the instrument for breaking and closing the circuit of the wires was rude, consisting of a slip of sheet copper with one end fastened to the table, while the other being loose moved like a spring, which was moved with the fingers to break and close the circuit. To remedy this rude contrivance, Avery, p. 433, states:

> I invented and made a finger key with a spring and fulcrum for breaking and closing the circuit of conductors, the construction

of which was my invention, the finger knob was made of ivory or pearl for insulating the key to prevent the electric shock.

He here produced a drawing of the finger key, which he invented, and which is similar to the same now in use in all the Morse offices. Mr. Avery at another place says,

> The introduction of the spiral spring was my own invention and suggestion.

He states, page 434, that he knew nothing of the invention of the local circuit until November 1843. But he says:

> On referring to the patent of Cooke and Wheatstone of 1837, and of Edward Davy of 1838, I find the local circuit as complete in form and principle, as in the patent of Morse at the present day.

Meaning Morse's patent of 1846, and reissue of 1848. Mr. Avery says that it was at the Railroad House in Baltimore, in July 1843, that he invented the finger key, or lever, as described, for breaking and closing the circuit. This Morse claims in his subsequent patents and reissues as his own contrivance.

Professor Gibbs, of the Free Academy New York, sustains, at page 437 for defence, the testimony of the previous witnesses. He states that Steinheil was the first to discover that the earth could be used as a part of a galvanic circuit. This he made by his attempts to employ the rails on a railroad to conduct currents of electricity.

At p. 438, he, with the foregoing witnesses, contends that Cooke and Wheatstone's patent of 1837 does contain the principle of an apparatus for sounding alarms, and the invention of the local circuit afterwards claimed by Morse, and especially in his reissue and patent of 1846. He says:

> Morse's recording instrument could not be worked at great distances without this (local circuit), or some precisely analogous contrivance, but Wheatstone's apparatus might record signals at very great distances, without the use of a local circuit, since it is more delicate than Morse's, and requires a current of less force to work it.

At p. 439 he gives very clear ideas of the difference between the branch circuit used by Bain, and the local circuit patented by Cooke and Wheatstone, and used by Morse:

Local Circuits—Relay and Receiving Magnets

> In Wheatstone's local circuit (like Morse's), an independent local battery is employed, which is wholly inactive until called into operation by the action of the main circuit, and which forms in itself a separate and short circuit entirely outside of the main circuit. In the branch circuit, used by defendants (Bain & Co.), a galvanic battery is made to form a portion of the main circuit at the station where signals are to be recorded; in order to record the signals transmitted, a short local or recording circuit is introduced, and the current is made to pass through the main circuit and recording circuit instrument.

He goes on to state, as others have done, that Bain's branch circuit is not only different from Cooke and Wheatstone's or Morse's local circuit, but is used for a different operation, viz., to decompose the metallic point in contact with paper moistened with the solution of the prussiate of potash, so as to produce chemical action, alone, in making signs; while Cooke and Wheatstone's local circuit is used by Morse, alone, to increase the mechanical force of his instrument. One acts chemically, and the other mechanically. He further says:

> In the local circuit arrangement of Wheatstone, or Morse, two independent galvanic batteries must be employed; in the branch circuit of the defendants, a single battery may be used to work both the main and the recording circuit, and I have seen one so used between New York and Philadelphia. (See p. 439.)

Again at p. 441:

> I have read and examined the specification of infringements referred to. I consider Mr. Morse entitled to nothing else than the precise combination which he uses; and as the defendants use a different combination, I do not consider them infringing upon Morse's just claims.

On p. 451, Professor Gibbs shows that Morse makes claims to matters in the reissue of his patent of 1846, made in 1848, not embraced in the original. Morse in his reissue of his 1840 patent, in 1846, claims the use of electromagnetic motion for telegraph purposes. Dr Bacon, instructor of chemistry in the Boylston Medical School of Boston, sustains the evidence of the previous witnesses, as regards dots, local circuits, receiving magnets, &c.; and at pp. 359, 360, states that:

The motive power of the electric current was used in Steinheil's telegraph, before named, to record signals of intelligence.

B. A. Gould, before referred to, at p. 83 for defence, says:

A magnet, such as Morse uses for a receiving magnet, is no more nor less than an electromagnet as already defined. Cook and Wheatstone invented the receiving magnet, which operated by deflection.

Page 89, Professor Jas. B. Rogers states that

The local circuit operates electromechanically, the branch circuit electrochemically.

Having shown by such a mass of evidence, that Morse, in the first place, was not the first to invent and use the local circuit; and that, even were it so, Bain's branch circuit forms no sort of interference with it, we proceed, in the next place, to examine the chemical telegraph, as the last contested point at issue of importance.

7
THE CHEMICAL TELEGRAPH

We come now to consider the chief remaining point at issue between Morse and Bain, viz., the Chemical Telegraph.

To settle this will require less space than was occupied in respect to the previous heads of the subject. For it will be seen from the testimony adduced, that Morse has nothing to claim in his chemical telegraph, that can by any possibility be considered as interfered with by Bain's patent.

Dyar, in his experiments on Long Island in 1826, claims to have been the first to try a chemical telegraph.

Edward Davy, of England, invented a recording chemical telegraph, which he patented in 1839. (See p. 44, *Evidence for Defence*.)

In that patent he makes the following claim: "Secondly, I claim the employment of suitably prepared fabrics for receiving marks by the action of electric currents, for recording telegraph signals, signs, or communications, whether the same be used with the apparatus above described or otherwise".

The "suitably prepared fabric" spoken of in this claim, referred to paper or cloth, moistened with solutions of salts, such as the hydriodate of potash, or the prussiate of potash, or other decomposable salts, by the action of the electric fluid, which would produce a black or dark colour when the fluid passed through it, and leave the paper unchanged when the fluid was cut off, or the circuit was broken.

Mr. Bain improved upon this patent of Davy's, both in respect to the mechanical apparatus contrived in connection with it, and in preparing the paper, with improved chemical compounds, which rendered its discoloration, by the action of the electric fluid, more delicate and certain. For these improvements he obtained a patent in England, which was enrolled June 12th, 1847; and in 1848 he obtained a patent in the United States. In 1848 Morse filed a caveat for a chemical telegraph, and opposed Bain's application. Commissioner Burke, as stated, decided in favour of Morse, on the ground that (contrary to all previous practice, in both this country and England) an English patent did not date from its enrolment. Bain made an appeal to Judge Cranch, who decided that he was entitled to a patent. While the case was before the Judge, the Bain party charge that Morse's specification, being discovered to contain some defect in some part of it, was amended at the Patent Office by the interlining of some words.

George Mathiot, connected with the United States Coast Survey, at p. 211 of *Evidence for Defence*, states

> I have read a description of such an electric telegraph (as that claimed by Morse), invented by Edward Davy, and patented in England in 1838. The arrangement is very similar to the telegraph of respondent's (Bain's), in this case.

Professor Augustus A. Hays, of Boston, states, at p. 343 of *Evidence for Defence*, that:

> E. Davy, in the patent above alluded to, uses prepared fabrics; and it is worthy of remark that the same chemical agent, viz. iodide of potassium, which Davy uses and prefers, is stated by Morse in his patent of 1st May 1849, to be the best of the substances he has mentioned.

It must be remembered that Morse's patent, in which he claims the use of this salt, was from nine to ten years subsequent to the date of Davy's patent.

At p. 315 *same evidence*, Professor Charles T. Jackson, of Boston, says:

> I have seen the model and patent of the chemical telegraph of Morse. It seems to me to be founded entirely upon the ideas received from me by Morse. My idea, as announced to him, was to produce marks upon paper properly prepared, by the electric current, and to interpret those marks by numbers. The marking part of this apparatus in this patent is the same in principle as that described in the patent of Davy.

Professor Eben Norton Horseford, Rumford Professor of Science applied to the Arts, and Director of the Chemical Laboratory in the Lawrence Scientific Department in Harvard University, shows in his evidence for defence, pp. 352 and 353, that the mere employment of different kinds of salts alone in solution, applied to paper, as proposed in Morse's patent, cannot be used successfully to make reliable marks, except in one or two instances, and then not so well as the same thing is done in Bain's arrangement, by a different process. Morse proposed to decompose a number of different kinds of salts which paper had imbibed, and to cause marks, or dots and lines, at the point of contact with a metallic point, through which the electric fluid was conducted. Professor Horseford shows, that many of the salts proposed to be used by Morse were utterly worthless; and Mathiot states the best named by him had been previously invented by Davy.

In Bain's patent and plan of working his machine, there is one essential point which renders it entirely distinct from Morse's, and proves that it in no respect interferes, in the slightest degree, with Morse's claims to a chemical telegraph, even had not Bain and Davy been in advance of him. That is, Bain wets paper with a solution of the prussiate of potash, places it on a revolving disk, and in contact with a fine point of iron or copper wire, through which the electricity is passed or checked at will. Now he does not use the electric fluid, thus passed from a battery, to decompose the *salts*, but to decompose the *metal*, or *iron* point. To enable him to do this, he adds nitric acid to the moistened paper, or solution of prussiate of potash. When the electric fluid is passed, this acid decomposes the *metallic point*, converting it into an oxide of iron or copper; and if the former, this oxide of iron converts, at its point of contact, the prussiate of potash into Prussian blue, and causes quite plain permanent blue or black spots or lines to be made. If copper wire be employed, the oxide of copper causes red dots or spots to be made. Bain prefers the use of the iron. Morse in his chemical patent makes no claim to any process of this kind whatever. Professor Horseford, referred to above, states at p. 354 for defence, that:

> Bain employs electricity, not to decompose salts, but to dissolve iron or copper, the blue or red stain resulting being the effect of the subsequent action of chemical affinity, unaided by electricity.

There is, therefore, no infringement upon the patent of Morse in the practical telegraphing by the process of Bain.

Morse, in his chemical patent, employed electricity to make coloured marks by the decomposition of salts.

Professor Jno. Bacon, M. D., of Boston, instructor of chemistry in the Medical School, sustains the evidence of Horseford, and states furthermore, that he did not then consider Morse's chemical telegraph available, (see p. 362 *defendant's evidence.*) He also sustains other witness at p. 362, regarding the use of dots, local circuits, receiving and relay batteries, &c. At p. 366 he says, speaking of Bain's patent:

> The electric current is not used to decompose the salt (prussiate of potash), but to cause the oxidation of the end of the steel or copper wire resting upon the chemically prepared material and solution of it in the acid. By this means a salt is produced which reacts chemically with the ferro-cyanide of iron (Prussian blue) or ferro-cyanide of copper (prussiate of copper). The marks obtained are very distinct and permanent, and require only a feeble electric current to produce them, especially if a steel wire is used.

Professor Gibbs sustains the evidence of the preceding witnesses, and at p. 441, *evidence for defence*, says:

> I have read and examined the specification of infringements referred to, I consider Mr. Morse entitled to nothing else than the precise combination which he uses; and as the defendants use a different combination, I do not consider them as infringing upon said Morse's claims.

Professor James B. Rogers, of the University of Pennsylvania, and successor to Professor Hare, sustains the evidence of the foregoing witnesses.

At pp. 91, 92, he gives a very clear analysis of the chemical telegraph of Bain. He shows that:

> The prussiate of potash is not decomposable by a current of electricity. But when nitric acid is added, to paper saturated with its solution, the electricity at the moment of passing disengages a portion of oxygen from the acid and water, which unites with the steel points producing the oxide of iron which is dissolved in the remaining acid, forming a nitrate of iron,

upon which the prussiate of potash acts chemically, producing the Prussian blue marks. So that the prussiate of potash is not decomposed by the current of electricity; but by the chemical action of a substance dependent for its formation upon the nature of the wire point being iron. This decomposition of metal by freed oxygen, by means of an electric current, forms no part of Morse's patent.

Other evidence could be multiplied all going to prove, beyond all cavil and doubt, that Bain's chemical telegraph in no way interferes with anything Morse has ever proposed in the same line.

FINGER KEYS.

Although Mr. Avery claims to have invented the finger key in 1844, and which is substantially the same as that used in the Morse offices at this time, yet it was in evidence that finger keys were in use long prior to either the claim of Avery or Morse. In allusion to the finger key claimed by Morse, Professor George Mathiot at page 211 *for defence* says:

> Such a finger key was in actual use in the German telegraph in 1837, while Morse was trying to operate with port-rules and type. It is also described in the patent granted to Cooke and Wheatstone by the United States prior to the patent granted to Morse, and patented in England in 1837, to the best of my knowledge. It is also in the patent granted to Edward Davy in England in 1838.

Dr Wm. E. Channing, at p. 381 *for evidence*, says:

> The signal key was one of the earliest and most obvious contrivances, in connection with the electro-magnetic telegraph. Ampère described the use of the key in 1820, and it has since been so generally employed that it will not be necessary to trace it further. See *Comptes Rendus* of the French Academy, Paris, 1838, p. 81; also *Annales de Chemie et de Physique*, Paris, 1820, vol. xv., p. 73.

Nearly all the witnesses referred to in what we have previously written, were also examined in reference to harmony or agreement between Morse's original patents and his reissues; and they generally concurred in the opinion, that there was a material discrepancy in some points; that terms were introduced not found in the originals, and that they disagreed in other respects. See evidence of Professor Gibbs, p. 451; also pp. 448 and 452, *for defence.*

In the evidence for the defence there exists a large amount of testimony, given by scientific men sustaining that from which extracts have been given; and which the limits of our review prevents our transcribing.

On looking over the evidence on both sides, we find the defence sustained by fourteen to fifteen men distinguished for their scientific knowledge, embracing professors and teachers of chemistry (electricity), with collateral arts and sciences. The list includes such men as Professor Joseph Henry, Professor C. T. Jackson, Wm. E. Channing, Professors Horseford, Gould, Benedict, Bigelow, Sears C. Walker, George Mathiot, O. W. Gibbs, Ruel Keith, A. A. Hays, Coxe, James B. Rogers, and others.

It is but seldom in a court of justice, we imagine, that such an amount of testimony as has been submitted—obtained from such a number of men separately examined and mostly out of court—has so uniformly agreed. They have all harmonised and sustained each other in every material fact; while for the complainants, or plaintiffs, only three men, of known scientific celebrity, gave testimony in opposition to that of the other side. These three men were, first, Dr Page, who was an examiner in the Patent Office, and passed Morse's patents and reissue in 1846, and Professors Renwick and Chilton. Included in the defence for plaintiffs, was a large amount of testimony taken from Professor Morse's relatives, including his own deposition, besides the testimony of a number of employees, &c., in telegraph offices.

We have seen from the analysis given, that Morse was not entitled to the exclusive use of dots, the local circuit, receiving or relay magnets; that signal keys and clockwork had been used before him; that he could not claim the exclusive use of electromagnetic motion, nor of the magnets invented by Henry; that he had made no discoveries in electricity or galvanism, or in magnetism or electromagnetism, or in anything appertaining to electricity in any respect whatever; that Davy had superseded him in a chemical telegraph, and that Bain's chemical telegraph was an entirely different thing from that claimed by Morse. Again it may be asked, what did Morse invent? Professor Gibbs replies to this question at p. 441 *evidence for defence*:

> I have read and examined the infringements referred to. I consider Mr. Morse is entitled to nothing else than the precise combination which he uses; and as the defendants use a different combination, I do not consider them as infringing upon said Morse's claims.

Mr. Simeon Borden, Civil Engineer, &c., states at p. 477 *for defence*, that trains of clockwork, similar to that employed by Morse, were in use before "his (Morse's) birth." As to the real character of what Morse is entitled to, it is summed up as follows by Mr. Borden at p. 479, and repeated on his cross-examination at p. 492. See *evidence for defence*.

- For industry and perseverance in the arrangement of an alphabet.
- For attaching a scratching point into the lever of an armature.
- For scoring the cylinder, over which the fillet of paper moves.
- For a combination of chemicals.

The last, however, Mr. Borden declared were not, and could not be successfully introduced in practice, and hence did not infringe Bain's patent.

8
DATES AT WHICH THE CHIEF TELEGRAPH LINES
IN THE UNITED STATES HAVE BEEN BUILT AND PUT IN OPERATION

1844 In this year the first line of electric telegraph was built in the United States, and extended from Washington to Baltimore about 40 miles, by S. F. B. Morse and his associates, Congress having made a grant of thirty thousand dollars to enable them to put it in operation.

1845 This year another line was opened between New York and Philadelphia, and Wilmington, by a company organised as the New York and Washington Telegraph Company, of which Amos Kendall was first president, Mr. Hart of Philadelphia secretary, and Thomas Clark of New York treasurer. Mr. Kendall was succeeded in 1850 by Mr. French, and he in 1851-52, by William Swain of the *Philadelphia Ledger*.

1846 Early in 1846 their line was completed by filling the link from Wilmington to Baltimore. For some time the news of battles in Mexico was brought by mail or express to Wilmington, and thence telegraphed to New York.

1846 In the spring of this year, a line was opened from Albany to Buffalo, under a company of which Theodore S. Faxton of Utica was elected president.

1847 June 22nd this line between Buffalo and New York was rendered complete, by opening the line between New York and Albany.

1846 A line was first opened between New York and Boston, under the management of F. O. J. Smith, with whom some of the Boston papers became offended at his management, and nicknamed him "Fog Smith," by which he became well known. He had purchased one fourth interest in Morse's patent while a member of Congress, between the years of

1838 and 1844. He was at one period Chairman of the Committee of Commerce, which reported the appropriation bill of $30,000 in favour of Morse, and which ultimately passed. He built the line in conjunction with a company of subscribers, and in virtue of a contract with Kendall, Morse, and Vail, co-proprietors of the patent. Smith finally purchased a sufficient number of shares, to give him a majority of the stock. He afterwards managed the line as he pleased, regardless of the views or wishes of many leading newspapers, with whom he became involved in bitter quarrels, and which resulted in the erection and encouragement of opposition lines under House's and Bain's patents, from New York to Boston, and of another line from Boston to Portland, where Bain's line joined on to the Portland and St. John's line. From this place, another line was continued by a provincial company to Halifax, and by which the New York Associated Press have received news brought by steamers to the latter town, without using Smith's line. In 1847 Smith built a line himself from Boston to Portland, and owned it as his exclusive property.

1846/47 Henry O'Reilly, Esq., a gentleman well known for his talents, perseverance, and energy of character, put in operation what he termed his Atlantic and Mississippi lines; the first of which extended from Philadelphia to Pittsburgh, and which was continued to Cincinnati. From this latter point other lines were built in 1847 and 1848, under his direction, extending from Cincinnati to St. Louis; and from thence lines in other directions, one of which reached Galena, in Illinois, and in its course interlocked with an extensive interlacing of Lake lines. He also pushed on a line from Louisville, south, to Nashville; and from thence a branch to Memphis, with a main line through Mississippi to New Orleans. More lines of telegraph have been erected under his direction than by any other person in the world.

A misunderstanding grew up between the Morse patentees

Dates at which Telegraph Lines built and put in Operation

and Mr. O'Reilly, with regard to the contract he had made with them for the use of their instruments on his lines. Mr. O'Reilly contended that they had violated, or refused to comply with their agreement. We have not room to give any details of the dispute. The result was, that Mr. O'Reilly, on his Louisville and Nashville line, introduced a new instrument, invented by Zook & Barnes. A lawsuit followed, as previously stated, and the case was carried from the District Court of Kentucky, by Mr. O'Reilly, to the Supreme Court, where it still rests, the decision in the District Court having been unfavourable to his cause. He soon after substituted the use of the Bain instruments in the same sections of country, and so far we have heard of no further suits regarding them in that quarter.

1847 A line was constructed by E. Cornell, Esq., of Ithica, N. Y., from Troy, N. Y., to Montreal, Canada, for a company organised for the purpose. It was in the same year that a line was opened from Portland to St. John's, N. B. It was also in this year, and early in the following summer, that Quebec, Montreal, and Toronto were put in communication. Lines were also built to connect Oswego with Syracuse, and Buffalo with Toronto in Canada, and Erie in Pennsylvania.

1848. In this year Mr. E. Cornell also built and opened a line extending from New York to Lake Erie, following chiefly the track of the Erie Railroad, and which was finally joined to lines erected by Mr. Speed and others in 1847 and 1848, extending to Erie, Cleveland, Toledo, Monroe, Detroit, and thence around the southern shore to Chicago, Milwaukie, &c., where in many cases they were interwoven with the O'Reilly, Lake, and Mississippi lines, and from which have since diverged others to Pittsburgh and Cincinnati. A company was organized in 1846, to build a line from Washington to New Orleans, and intermediate places along the seaboard. This was put in operation as far as Petersburg, Va., in 1847, and opened the entire distance to New Orleans in 1848. Elam Alexander, Esq., of Macon,

Georgia, became the president of the line. This forms the longest continuous line under the control of a single company. It connects the principal seaboard commercial cities of the South with New York. The company have extended a wire from Washington to New York, where they have established an office.

HOUSE LINES

1847-1848 A line was built to be worked by Royal E. House's printing telegraph instruments between New York and Philadelphia. The line was built for a company by G. E. Downing, Esq., of the latter city.

1849 The same party built a line to be worked by the House instruments between New York and Boston, which Mr. Downing presided over, and of which Mr. Sturges, of Boston, was manager. Changes in the management of both lines, however, have since been effected, and the proprietorship of the New York and Philadelphia line has changed hands.

1850 A House line was built to connect New York and Buffalo with intermediate places, by plans of insulation supplied by Mr. House, and executed by Mr. Edson and others. This, taken altogether, forms one of the most substantial and best-built lines to be seen in the country.

A House line has since been built and put in operation between Buffalo and Cincinnati.

LINES WORKED BY BAIN'S PATENT

1850 A line in this year was constructed and opened, to be worked by Bain's instruments, between New York and Washington. The line was built under the superintendence of Henry D. Rogers, Esq. The company who constructed this line were sued by the representatives of the Morse patent, and a trial had before Judge Kane, of Philadelphia, who decided unfavourably to the claimants under Bain's. An appeal was taken to the Supreme Court. Soon after, however, the Bain company sold out their line to the Morse company.

Dates at which Telegraph Lines built and put in Operation

1850 Another Bain line was constructed and opened between New York and Boston, called the "*Merchants' Line*" of which Marshall Lefferts, Esq., became president. This line was remarkably well constructed, and has been at all times one of the best managed in the United States.

1851 Another Bain line was built and put in operation between New York and Buffalo, which was constructed under the supervision of Henry O'Reilly, for a company. Mr. Lefferts also became the president of this line, and Mr. McKinley the superintendent of both lines.

1851 An extension of the Bain line was built from Boston to Portland, which connected with the Portland, St. John's, and Halifax lines.

About the same time Professor Benedict, of Vermont, constructed a Bain line, which connected Boston with the Montreal lines, passing through Vermont.

Several other short lines have also been constructed in different parts of the country, forming an extensive network of wires of a greater aggregate length than exists in any other part of the world.

9
NEW PROJECTED TELEGRAPH LINES
TO FACILITATE THE TRANSMISSION OF NEWS BETWEEN THE OLD AND THE NEW WORLDS, AND TO UNITE IN COMMUNICATION THE ATLANTIC WITH THE PACIFIC

The importance of speedy communication between the Atlantic and Pacific is quite obvious. While Mr. Whitney has perseveringly pressed forward his scheme for the construction of a railroad, Mr. Henry O'Reilly, who constructed so many telegraphs in the West, (with a main stretch connecting Philadelphia and St. Louis)—estimated by himself at about 7,000 miles—has followed up his plan for building a great national line, leading from the Mississippi to California.

Mr. O'Reilly presented a memorial on the subject at a previous session of Congress, which has been renewed at the present session (March 1852). This renewal is dated New York, December 1851; and was presented in the Senate by the Hon. Mr. Douglas, of Illinois.

To enable the reader to understand the plan proposed by the memorialist, we give the following extracts from the document. He heads his memorial thus:

> Telegraphic and Letter Mail Communication with the Pacific—Including the Protection of Emigrants and the Formation of Settlements along the Route through Nebraska, Deseret, California, and Oregon, with branches to New Mexico, &c.—and facilitating the correspondence across the American Continent between Europe, China, Hawaii, Australia, &c.

He afterwards proceeds to say:

> The fact that the undersigned SOLICITS NEITHER MONEY NOR FAVOR from the Federal Government, may at least free this Memorial from some of the difficulties usually connected with individual applications for governmental attention. The undersigned ASKS NOTHING FROM THAT GOVERNMENT which should not be shared in common with all citizens whose business requires PROTECTION OF LIFE AND PROPERTY ACROSS THE PUBLIC DOMAIN. Having been sustained by PUBLIC CONFIDENCE, and

The Electric Telegraph in the United States

not by any Governmental assistance, from the commencement of Telegraphing in America down to the present period, he prefers to continue that reliance upon his fellow-citizens, individually—being well assured of adequate support in this enterprise from energetic capitalists and business-men—rather than solicit from Government any assistance which may not be commonly enjoyed by all persons who embark their lives and property in Telegraphic or other enterprises through the Public Domain, between Missouri and California.

* * * * * * *

The proposition is substantially to the following effect:

That Congress shall pass a law, providing, that instead of establishing forts, with hundreds of men at long intervals apart, the troops designed for protecting the route shall be distributed in a manner better calculated to promote that and other important objects on the principal route through the Public Domain—namely, by stationing parties of TWENTY DRAGOONS AT STOCKADES TWENTY MILES APART:

And providing, also, that two or three soldiers shall ride daily, each way, from each stockade, so as to transport a DAILY EXPRESS LETTER-MAIL ACROSS THE CONTINENT; while at the same time protecting and comforting the emigrants and settlers; and thus incidentally furnishing all the protection which the undersigned invokes as a necessary preliminary for completing the comparatively short link of Telegraph between Missouri and California—short, comparatively, as contrasted with the seven thousand miles of Telegraph constructed under his arrangements in the First Division of the Atlantic and Pacific Telegraph.

* * * * * * *

With such a System promptly and liberally carried into effect, the undersigned does not hesitate to repeat the prediction, that, within two years, *at farthest*, the EUROPEAN NEWS may be PUBLISHED on the American shores of the PACIFIC OCEAN WITHIN ONE WEEK from the sailing of the steamers on the "shortened route" between the OLD WORLD and the NEW.

Whether Congress shall see fit to grant the memorial, or whether such a scheme is really practicable at the present time,

New projected Telegraph Lines

is immaterial.

It is certain that, as soon as the Indians are brought into peaceable subjection, and a good highway is established from Missouri to California, Electrical Telegraph lines will be built over the entire route. In the present rapid growth of the country, it is probable that within ten years New York will be in telegraphic communication with San Francisco.

Another enterprise of scarcely less importance, and with every prospect of early consummation, has lately been started.

It is said steamers can make ordinary passages between Cape Race, Newfoundland, and Galway, Ireland, in five days.

A company has been formed to construct a line of telegraph from Halifax, N. S., to Cape Race, with a capital of £100,000 (near $500,000).

> The Engineer of the Company is Mr. F. N. Gisborne, late General Superintendent of the Eastern lines of telegraph in the British Provinces, who surveyed the route last year, and will leave in a few days for Europe, to make contracts for the submarine wire. This projected line of telegraph completed, New York will be brought within five days of London. It is yet undecided whether to run the line to Cape North, at the Northern extremity of Nova Scotia, or to Prince Edward's Island. Should the former route be chosen, but forty-eight miles of submarine wire will be required; if the latter, one hundred and thirty; but the adoption of this will reduce the distance three hundred miles, and the line will pass through a thickly populated country, from which considerable local support will be derived. The Company is guaranteed the exclusive right to telegraph across Newfoundland for thirty years, and is granted a bonus of thirty square miles of land and $30,000. It is expected that the whole will be completed and in operation in six months from the present time." See the *Journal of Commerce* of the 9th April 1852.

Another project is on foot in England, to cross the St. George's Channel by means of a submarine telegraph, as has already been accomplished between Dover and Calais, across the English Channel. From the St. George's Channel, it will be an easy matter to extend the wires to Galway, on the west coast of Ireland.

When these enterprises come into operation, including communication between New York and San Francisco, what will then be the position of the civilised world in regard to telegraphic intercourse? All the chief towns of this vast continent will be within five days communication with all the capitals and principal cities and towns of Europe.

Steamships can make regular voyages between San Francisco and the Chinese ports of Canton and Shanghai, in from 22 to 23 days, and probably in 25 or 30 days to Australia. Hence the time will soon arrive, when New York will be brought within from 22 to 23 days' communication with China, and all the cities of Europe can also communicate with China in from 27 to 28 days!

Thus, a merchant in Liverpool or New York, after having dispatched his ship for a month or more, with a cargo for China, can, on a change of the markets in either place in the articles shipped, or in the value of teas or silks to be purchased, forward fresh instructions to his consignee or supercargo, which will reach the Chinese ports in advance of his vessel. In this point of view, the value of such speedy communication becomes incalculable.

A recent proposition has been started to connect the island of Cuba with the peninsula of Florida, by a submarine telegraph. It is said parties in Cuba have obtained a grant for this purpose, and propose organising a company, embracing parties in the United States, to carry it into effect. We imagine this enterprise, from the great depth of the intervening sea, and the coral irregularities of the shores, will be attended with considerable difficulties, though probably not of an insurmountable character.

A telegraph line has been authorised by the Mexican government, to connect Vera Cruz with the capital, and to be continued from thence to Acapulco, on the Pacific.

It is said the section between the city and Vera Cruz has been for some time in the course of erection, and is soon to be completed and put in operation. Another line was projected. to connect the city with Matamoras, thence extending through Texas to New Orleans. The erection of such a line must be considered rather remote.

10
STATISTICS OF TELEGRAPHS IN THE UNITED STATES
PLAN OF ERECTING LINES
METHOD AND EXPENSE OF WORKING THEM

The length of telegraph lines built and in operation in the United States and Canada, is estimated at from 12,000 to 15,000 miles. The most distant points in communication are Halifax, N. S., and Quebec with New Orleans, near 3,000 miles intervening between them, following the circuitous routes of the wires. The towns and villages which are accommodated with telegraph stations amount to between 450 and 500. As there are competing lines, under different companies, between New York and other principal cities, many of the towns have two or three separate telegraph offices. New Orleans is connected with New York by two lines. The first passing south by the way of Washington, Richmond, Charleston, Savannah, Augusta, Macon and Columbus, Ga., and Montgomery and Mobile, to New Orleans. The other passes via Pittsburgh, Cincinnati, Louisville, Nashville, and thence through Mississippi to New Orleans. Each of these routes intersects with other lines, and give off lateral branches to many places not on the main routes. The distance traversed by either line from New York to New Orleans does not vary much from about 2,000 miles. Messages passing from one of these cities to the other have usually to be re-written four or five times at intermediate stations; though, by an improved method of magnetic connections, the sea-board line has, in good weather, transmitted communications direct between New York and Mobile, without intermediate re-writing, a distance of near 1,800 miles. By the Western or Cincinnati route to New Orleans, steamers' news handed in at 8 A.M. has reached New Orleans, and the effects produced on the market at that point returned to New York by 11 o'clock A.M. Short messages forwarded from New York have frequently beaten time in reaching St. Louis and New Orleans.

The Electric Telegraph in the United States

To illustrate the speed with which news is sometimes transmitted, we give an extract from the *New York Herald* on another page.

The dispatch referred to was one that we had received from our correspondent in Liverpool, and forwarded to the Merchants' Exchange in New Orleans. The plan of operations which we adopted was as follows. We requested our Liverpool friend to prepare a synopsis of commercial news, up to the moment of the departure of each steamer, and to make four copies of it on manifold paper. It was also condensed into a form ready for transmission the moment the steamer reached our port. We hired news boatmen to cruise down the harbour and watch for the steamers. As soon as one appeared at quarantine they would board her, and obtain our bag from the hands of one of the employees of the boat. This bag would contain the latest Liverpool and London papers. As soon as our boatmen obtained the package, they would make all possible speed for the city both by oars and sails. On landing at the docks, they would immediately fly to the telegraph office with manifold slips, one of which was invariably sent to New Orleans as stated.

On one occasion late in the afternoon, the *Asia* arrived at quarantine. A strong southerly breeze was blowing, and our newsmen set sail and reached the city in very quick time. The news was put into the telegraph office for New Orleans, it reached there and an answer was returned before the steamer came up.

Upon another occasion, when great anxiety prevailed regarding the safety of the *Atlantic*, the *Africa* arrived after dark off quarantine, and it was generally conceded that unless she brought some tidings of the missing ship, that she must be given up for lost. Again our newsmen brought the first intelligence received by the *Africa* to the city, that the "*Atlantic* was safe!" We immediately sent the news to all parts of the country, and also to Mr. E. K. Collins, the consignee of the *Atlantic,* who was then in Washington, and who afterwards informed us that our dispatch gave him the first intelligence of the *Atlantic's* safety. We had also the satisfaction of knowing that our news reached all parts of the country, before the *Africa* came up to the city.

We shall never forget the thrill of joy with which the news was everywhere received. Every public place down town, and newspaper office, was besieged by hundreds, if not thousands in this city, as soon as the *Africa* was telegraphed as being in the offing; and they awaited with breathless anxiety the result of her tidings regarding the *Atlantic*. At last the news came. It was read aloud to them—"*The Atlantic is safe!*" when there arose loud and enthusiastic shouts of joy. It flew from mouth to mouth—from one extremity of the city to the other—along the shipping—among the ship-yards and ship-builders—among those who had worked on the missing vessel. It flew abroad to the suburban towns. It became a theme of exultation at the hotels and theatres. In some of the latter, the managers came on the boards and announced to their auditors that "*The Atlantic is safe!*" which was followed by the rising of the whole audience to their feet, and giving the most deafening and enthusiastic applause. In our whole experience in telegraph reporting, we recollect no instance in which a piece of news gave such universal delight. No battle ever won in Mexico diffused greater satisfaction in New York than the safety of the noble ship *Atlantic*. The circumstance furnished evidence of the strong current of national feeling which was associated with the success of the "*Collins Line of Steamers*" and which have so nobly rewarded that proud national feeling, by distancing all competition.

We return to the extract regarding the *Europa's* news, from which we have digressed. Here it is:

> The *Europa* reached her wharf yesterday at 6½, A.M. Her news was at the O'Reilly Telegraph Office, 181 Broadway, somewhere before 7, previous to the office being opened. The Pittsburgh office got to work about 8 A.M. and the dispatch commenced going to New Orleans 10 minutes past 8, and was received and put up in the Exchange before 9 A.M. and the acknowledgment of its receipt as at foot reached the O'Reilly Telegraph Office, New York, at 11¼ — thus having travelled from New York to New Orleans and back in three hours and five minutes.
>
> O'Reilly's Telegraph Office,
New Orleans, May 8.
>
> *To Smith, Chief Operator, New York Office:*
> The foreign news per *Europa*, signed "Jones," was received here before 9 o'clock, A.M., New Orleans time.
>
> Signed "ZOOK, Chief Operator."

The Electric Telegraph in the United States

This news, in its transmission, as will appear by the following note from Mr. Baily, clerk in the O'Reilly Telegraph Office, was only re-written three times:

181 Broadway, May 8 P.M.

I have to state that the foreign news by the *Europa* was forwarded from this office at 8h. 10m., A.M. It was only rewritten at Pittsburgh, Louisville and Tuscumbia, before reaching New Orleans, where it was received and hung up in the Merchants' Exchange before 9 A.M., New Orleans time. The message acknowledging its receipt as above was received at this office at 11¼, A.M.

Signed "BAILY, Operator."

The distance between New York and New Orleans, following the track of the telegraph lines, is about 2,000 miles, and may somewhat exceed it.

One important function of the electric telegraph, although previously referred to, is not likely to be duly appreciated by many readers. The electric current not only conveys a message from one distant point to another, but, like a skilful letter-carrier, it can be made simultaneously to drop copies of the same at the intermediate stations through which it passes. This is done by putting the instruments at each station in the general circuit. To illustrate our meaning: The Congressional reports put in the office at Washington are usually received simultaneously in Baltimore, Philadelphia, and New York; and all that is necessary, at the intermediate stations, is for an operator or clerk to be present and receive the message as it is developed on paper by the instruments. Thus, also, the usual practice in supplying the press of Buffalo, about five hundred miles from New York, and at intermediate places, such as Albany, Utica, Syracuse, Rochester, &c., is to put the line in a continuous connection with the intervening instruments, forming parts of the grand circuit. Hence, the same commercial and general news is read off simultaneously at every office on the entire route. The news for the press in that direction is usually sent, within certain hours, twice a day, and charged for at so much per week for each paper on the line receiving it. This facility of taking drop copies at intermediate stations, however, causes important news to become known to a greater number of persons, and renders its concealment more difficult.

Owing to the want of experience, and limited command of capital, the first lines were very imperfectly constructed. The best modes of insulating wires on poles had not been devised. The timber employed was small and faulty. The consequence was that many of the early lines soon required expensive repairs, or had to be partially reconstructed. And up to the present period, few of them have been built in a manner to secure the greatest durability and perfection of insulation. The actual cost of the best of them has not exceeded $200 per mile. To build a line in the best manner to render it permanent and free from interruption, would probably cost from $400 to $500 per mile. Such a line would not exceed about one-twentieth the cost of a good railroad. A great central national line, conducted in the best manner, to connect New York and New Orleans in daily, speedy, and reliable communication, is much needed. Such an enterprise would be sufficiently national in its character, to deserve the liberal patronage of the government. Such a line, constructed so as to render interruption from any cause next to impossible, would pay well as an investment. The irregularity attending communication with New Orleans, and especially at certain seasons of the year, is so great, that a great many merchants, who would otherwise correspond regularly, are now deterred from doing so. At the present time, the various telegraph lines in operation are worked by the three instruments previously referred to, viz. Morse's, House's, and Bain's. The lines have been built and are owned by various joint-stock companies, estimated at between twenty and thirty.

The plan proceeded upon by Professor Morse and his associates, was, to dispose of the right to companies for half of the nett receipts. In other words, a line being estimated to actually cost $150,000, was assumed to be worth $300,000. The first sum being paid up by subscribers, they would receive stock for that amount, while an equal or duplicate amount of stock would be issued to the patentees. When dividends were earned, they would be divided on the $300,000 assumed capital. Mr. Bain acted on the same principle, only that he received one third of stock on the amount subscribed, and on minor routes even less. Mr. House in some cases proceeded on a similar plan, while in others, as in the New York and Philadelphia House line, he sold out his patent-right to the party who built it; and this party has since sold out to others.

In Prussia it has been determined to carry the principal telegraph lines underground in gutta percha tubes; and it is said that it can be effected at an expense of only about £40 (about $200) per mile. This estimate we believe to be far too low. There will be, no doubt, many difficulties experienced in attempting to work long lines underground, not the least of which will be in finding and repairing breaks when they occur.

In France, it is said, after much discussion and consultation, they have determined to carry the wires on poles.

In building a permanent line, the time may arrive when it will be found, in the long run, most economical to adopt erections of iron rods (or poles), secured in blocks of stone, and well braced. Such a line would stand as a permanent structure in all seasons, and might be made to last over a century.

At an early period, Professor Morse disposed of interest in his patent to other parties. Among others who became partners in his patent, were Mr. Amos Kendall, of Washington, and who also became his attorney, Mr. F. O. J. Smith, of Maine, Mr. Alfred Vail, of New Jersey, and Leonard D. Gale, of Washington. The latter party, we understand, has since parted with his interest.

In these arrangements, Professor Morse has only retained one-fourth part of his patent in his own right, and hence receives one-eighth of the net dividends of the principal lines worked under his patent claims. It is said that, even under this arrangement, his revenue has been quite large, it having placed him in very independent, if not wealthy circumstances.

As before stated, Mr. F. 0. J. Smith constructed and became chief owner of the Morse line from New York to Boston; and, we believe, also the Morse line thence to Portland, Maine, exclusively. The management of these two lines, by Mr. Smith, was such as to give great dissatisfaction to the press and to the public; and also to the other joint proprietors of the Morse patent, including Professor Morse himself.

Statistics, Plans and Methods of Working

The disagreement finally became so great that Messrs. Kendall and Morse sued Mr. Smith for large claims, alleged to be due under pre-existing contracts. The case is at present in fierce litigation.

The cost of constructing lines varies very materially according to the nature of the country through which they pass. The expense is greatly influenced by the price of labour, and intervening water courses and inlets, which have to be crossed.

To arrive at an approximation of the amount of capital invested in all the telegraph lines in the United States and Canada, let us assume their total length to be 12,000 miles. Exclusive of stock issued to patentees, the average cost of all the lines, including instruments, may be set down at about $175 to $200 per mile. Twelve thousand miles, at $200 per mile, makes an aggregate of $2,400,000. If we add one-third more for the use of patents, we shall have a representative capital of $3,200,000 for the entire cost of the present telegraph lines on this continent.

The amount of dividends declared by the different lines has varied very much. On some they have been very large; while on others, where repairs have been heavy, they have been small. The largest have been made on the Morse New York and Buffalo, and on the O'Reilly line from Philadelphia to Pittsburgh and Cincinnati. The heaviest have reached from 16 to 25 per cent, per annum, while some have been as low as 4 per cent.

The Washington lines have done well, and the earnings of the Morse line from New York to Washington have been large; but much of its income for the past year or two was spent in rebuilding the line, and keeping up communication across the Hudson river. The Washington and New Orleans Morse line has always done a good business when at work; but it has been very liable to interruption from the falling of trees in the Southern pine forest through which it passes, and from the frequency of thunder storms.

All the lines between New York and Boston have, as a general thing, done well. The Bain Eastern line has succeeded well, under excellent management; and being the only connecting link with the lines leading from Portland to Halifax, by which the New York associated press receive the foreign news brought by steamers to that port, adds materially to their other receipts. Like all other kinds of business, the relative earnings of different companies depend very much upon the skill and judgment with which they are managed.

Few of them ever publish to the world regular reports; and so far no telegraph stock has appeared on the brokers' books, nor has it been bought and sold at the stock exchange.

Stockholders would be benefited if telegraph companies, like other leading corporations, would make public annual, or semi-annual reports, by which outsiders could arrive at some conclusion with regard to the intrinsic value of the shares appertaining to each line.

The most frequent causes of interruption to telegraph communication, in this country, arises from storms and sleet in winter; and from thunderstorms in summer.

It is believed that interruption from atmospheric electricity can be prevented by extending a wire on top of the poles the whole length of the line, with ground connections, at suitable intervals, which would convey the aerial electricity to the earth, and leave the other wires beneath it free from interruption. All the companies between the principal places have two wires on the same poles, and in some cases three and four.

The superintendent of the Washington and New Orleans line discovered, while in the south during warm summer weather, however clear the atmosphere, that, as the sun rose and advanced to the zenith, the uppermost wire of the two employed, would become deranged by the influence of atmospheric electricity. He then reversed the wires, placing the lower wire at top, and the other beneath, and yet found this lower wire would work, while the upper one was interrupted. This fact has suggested to our mind, that if an upper wire were given up to atmospheric electricity, with ground connections, that all beneath it would be free of interruption from this cause.

At present, each company employs men, at suitable distances, to look after the condition of the wires. In the thickly populated country of the north, these guardsmen are placed at long intervals; it may be of forty, fifty, or one hundred miles apart. On lines passing through the dense southern forests, it has been found necessary to employ them at intervals of about every twenty miles, and to make it their duty to traverse and examine the lines frequently, and especially during or after storms. The expense of guarding and repairing lines varies materially, according to circumstances, and differs with each line; but forms the largest item withal.

11
EXPENSE OF BUILDING AND OPERATING THE LINES

The local expenses at the stations, or offices, vary according to circumstances or localities. A large expense arises from the use of acids, and the decomposition of metallic zinc, in the batteries. From some data gathered from inquiries made in the proper quarter, it is estimated that the zinc cups employed in the batteries average about twenty-five to thirty for every one hundred miles of telegraph wires throughout the United States. On routes where there are two or three separate telegraph lines, under different patents, the number of zinc cups required will be double and treble the quantity named; but on some single routes, with small stations, the quantity is, of course, much less. But, for the sake of calculation, we will assume that the average is thirty to the hundred miles. Now, assuming the entire length of lines to be 12,000 miles: to work them will require the use of 3,600 cups. Each zinc cup weighs from 2 to 2½ lbs. If we take the latter estimate, the weight of the 3,600 cups, will amount to 9,000 lbs of metallic zinc. These zinc cups undergo total decomposition, and have to be entirely replaced with new cups every six months; hence, for the year, the consumption of metallic zinc will be 18,000 lbs, which at 8 cents per lb, gives only an aggregate cost of $1,440 per annum. Some companies renew their cups more freely; and if we add the expense of transporting the cups, it is likely the whole consumption of metallic zinc, may reach $2,500 to $3,000 per annum.

The next heaviest expense connected with the working of the batteries, arises from the consumption of nitric acid. This is poured into the porcelain cups, in which are placed the slips of platina foil. Nitric acid, of best quality, costs about 11 cents per lb. It takes about 1 lb. to every eight porcelain cups, or about 12½ pounds to every 100 cups. Now, as we have shown that 3,600 zinc cups are required, hence, to work them, they require an equal number of porcelain cups. Thus, to fill them once, for operation, will require about 450 lbs. of nitric acid, which at 11 cents per lb., will cost $49.50. This acid loses its power and requires to be entirely renewed about twice a month in the main batteries, and daily in small local batteries, of only two or three cups each. Hence the consumption will cost, for all the lines, $1,188 per annum, which may be considered a low estimate. Taking wastage and expense of transportation into account, it is possible the amount may reach from $1,500 to $2,000 per annum.

The next expense is that of sulphuric acid; but as it is only used in small proportions, and differs in strength and quantity used by each battery, we cannot arrive at a probable estimate. As it only costs 6 or 8 cents per lb., and its value is trifling, in proportion to that of the nitric acid, we will pass it over. To each battery of 100 cups, six lbs. of quicksilver or mercury is employed, to rub over the zinc cups, for the purpose of causing them to resist decomposition by the action of the acid. The mercury costs about $1.25 per lb., and if six lbs. per 100 cups per annum be employed, it will cost $7.50 per 100 cups, or $370 per annum. If we add incidental expenses, the amount may reach $500 to $600 per annum.

The aggregate cost of materials consumed in working all the telegraph lines on the continent for one year, may be summed up about as follows, viz:

Metallic Zinc, say	$3,000
Nitric Acid	2,000
Mercury or Quicksilver	600
Breakage, Wastage, &c.	500
	$6,100

By this statement, it will be seen at what comparatively small expense of materials, the electric fluid is daily sped to all parts of this vast country. Further improvements will likely reduce the present cost of materials.

The offices in New York employ on an average four young men in each, as operators and clerks. Those acting as clerks are generally also capable of writing with the instruments. In country offices, or in places of small note, one or two operators are sufficient. The wages paid for their services differ in different offices. The chief operator receives the highest wages—varying, probably, from $1,000 to $1,200 per annum. Some companies also employ a person known as the superintendent of their lines, who has the immediate control and supervision of the whole. It is customary with most, if not all the offices, when the operators have reached the hour for closing, or have finished their day's work, if required by the press or other parties to keep open for a longer period, to charge those giving the order extra for their services. As two have to sit up in each office, the usual charge is 50 cents per hour for each person, or $1 per hour for each office. The offices in New York manage the delivery of their own messages. For this purpose they employ, on an average, about five boys each, for twelve offices, making an aggregate of about sixty boys. The wages paid these boys is from two to three cents for each despatch delivered, if below Canal Street, or within about a mile from the offices. If beyond that distance, or after night, the charge is 12½ cents for each despatch. If in distant parts of Brooklyn or Williamsburg, the charge is 25 cents. Each boy carries a small book or register. The envelope containing the message is endorsed with the address of the party to whom it is directed, and the name of the office at which it was received. The time at which it was received is inserted over the message, and within, the place from whence it came, with the date, &c.

When the boy delivers the despatch, the recipient is requested to write his name and residence in the boy's book, and the precise time at which he delivered it. These books must always be produced at the office when called for.

Besides the help previously referred to, many of the offices employ what may be termed a battery man, whose duty it is, every night, to remove the zinc cups from the acid cells or cups, and, after cleansing them in clear water, to set them by until they are required for use next day.

12
STATISTICAL AND OTHER INFORMATION
REGARDING THE OPERATION OF SEVERAL LEADING LINES. QUESTIONS AND ANSWERS FROM THE O'REILLY, BAIN, HOUSE AND MORSE NEW YORK AND BUFFALO LINES

QUESTIONS

1. What is the greatest distance from which electric messages are actually transmitted?
2. What are the tariffs of transmission?
3. What classes of dispatches are entitled to precedence?
4. What is the average number of words (accidents apart) which are transmitted along a single wire per minute? If the different telegraphs differ in their rate of transmission, state it?
5. At what rate can dispatches which arrive in telegraph cipher, such as Morse's and Bain's, be reduced to ordinary writing?
6. What quantity of telegraphic matter forms an average per day?
7. Give examples of days on which you receive extra quantities?
8. To what extent do you receive telegraphically the debates of Congress?
9. What convention exists between the New York journals for telegraphic news?
10. To what extent is the telegraph used for commercial correspondence? This would be best illustrated by the expense incurred by two or three of the greatest commercial houses.
11. Is the telegraph extensively used for social correspondence?
12. State the cases in which there are competing wires, and illustrate the effects of competition by examples?
13. Are interruptions frequent, arising from atmospheric electricity?
14. Or from other, and what causes?
15. How many breaks are there in the communications between the most distant stations, &c.?
16. Is it practicable for correspondents to keep the subject of their dispatches concealed from the employees of the telegraph, and is this object often, or ever practically attained?

The Electric Telegraph in the United States

ANSWERS TO THE FOREGOING QUESTIONS MADE BY MR. O'REILLY AT HIS TELEGRAPH LINE OFFICE, 180 BROADWAY.

1. The actual distance from which messages have been, and are now transmitted on this line, is 1,100 miles—from New York, to Louisville. To do this, it is found necessary to place two batteries in the circuit at distances of four hundred miles apart, for the purpose of renewing the electric current, part of which escapes from defective insulation and atmospheric causes. There is no doubt but that, in a more advanced stage of telegraphing—which may be but a short time hence—that New Orleans and New York will be placed in instantaneous communication with each other. To enable this to take place, requires, in the first place, a line substantially built and thoroughly insulated. It may be remarked, that it is but two years since, when to telegraph three hundred miles on a single or unbroken circuit, was considered a feat; now, from improvements made since then in telegraphs, we can send over one thousand one hundred miles—easier than we could three hundred at that time. In our Cincinnati office, two years ago, and up till very lately, they used a separate battery for each line. From a series of experiments made, one single battery, of no greater strength than those formerly used, now works eight distinct and separate lines, with no apparent diminution of strength, and at a great saving of expense to the office.
2. The tariffs depend, in a great measure, on the competition which the lines have to meet. From New York to Pittsburgh, the rate is sixty cents for ten words, distance four hundred miles; ten words to Cincinnati, distance seven hundred and eighty miles, is but seventy-five. Owing to the competition existing among four or five lines that connect Cincinnati with New York, by adding the rates and distances on three or four of the lines, it would give, as an average:—That ten words of a message (the address and signature of sender and recipient are not counted in telegraphic messages), are transmitted, in long distances of over four hundred miles, at the rate of one cent; three hundred miles, about six mills; two hundred miles,

about five mills; one hundred miles, about four mills; and lesser distances in proportion. †

3. Classes of messages entitled to precedence, are government messages, and messages for the furtherance of justice in detection of criminals, &c.; then death messages, which includes cases of sickness when the presence of a party is requested by the sick and dying. Important press news comes next; if not of extraordinary interest, it takes its turn with the mercantile messages.
4. Average number of words may be stated at twenty to twenty-three per minute; a higher rate could be obtained, but as nearly all operators copy from their instruments and reduce messages from ordinary writing, the above is considered rapid enough— as an expert operator can indent his Morse characters on his register faster than most men can write with a pen or pencil.
5. The messages, as quick as they are written by an operator at the extremity, are copied at the other extremity by the receiving operator on a printed slip of paper, prepared for that purpose; then passed from the operating to the receiving room, enveloped and sent out for delivery. So that no delay need take place, if each one attends to his duty.
6. That is a question that cannot well be answered here as regards this line, as we are at the extreme point, and we are not one of the feeders to the line. New York, Philadelphia, and Baltimore messages for the West and South, all pass through the Pittsburgh office. From the last report of the line from Louisville to Pittsburgh, presented to shareholders by the superintendent, the following statistics are stated. It may be mentioned that this section is 450 miles long, distinct as a property, but under one management.

Statistics of the year 1850 — Pittsburgh and Cincinnati Telegraph Line.

Number of words transmitted ···················· 3,602,760
Number of dispatches recorded ······················ 364,559
These are exclusive of free matter, necessarily large at all times.
Average hours of labour, fourteen hours per day.

---

† Editor's note: A mill is a notional unit equivalent to 1/1000 of a United States dollar (a one-hundredth of a dime or a tenth of a cent).

The recorded dispatches for 1850, on the paper of the registering Instrument, covers a length of 1,704½ miles.

Cash receipts at the different offices of the Pittsburgh, Cincinnati, and Louisville Telegraph Company, during the year 1850.

Louisville	$22,000.08
Madison	2,155.99
Lawrenceburg	292.60
Cincinnati	18,970.97
Dayton	2,727.55
Springfield	631.37
Columbus	3,403.49
Zanesville	1,628.36
N. Washington	72.37
Wheeling	2,525.71
Steubenville	878.08
Pittsburgh	17,992.17
Total receipts for 1850	$73,278.74

Record of Dispatches—1850.

Pittsburgh	73,900
Steubenville	5,020
Wheeling	12,100
N. Washington	242
Zanesville	5,079
Columbus	12,885
Springfield	2,700
Dayton	9,968
Cincinnati	156,000
Lawrenceburg	1,000
Madison	10,325
Louisville	74,660
Total number of entries for 1850	363,879

This table, of course, shows a double entry—one at the office whence sent, and the other where received; yet the record as given, showing as it does all the business created by each office named, also shows thereby their intrinsic value.

7. and 8. We do not receive Congressional reports. We are not in communication with Washington.

Questions and Answers—O'Reilly, Bain, House and Morse Lines

9. The rates for press-matter on this line are as follows:

	Per Word.
New York to Pittsburgh 3	cents.
Philadelphia 1½	"
Other dispatches between that and New York 3	"
New York to Wheeling 3	"
Zanesville, Ohio 3	"
Columbus, Ohio 3	"
Dayton, Ohio 4	"
Madison, Ind. 4	"
Louisville, Ky. 5	"
Nashville, Tenn 6	"
Memphis, Tenn. 6	"
Jackson, Miss 7	"
Vicksburg, Miss 8	"
New Orleans, La. 8	"
St. Louis, Mo. 6	"
Lancaster, Penn 2	"
Harrisburg, Penn 2	"
Philadelphia to Pittsburgh 1½	"
Wheeling 2	"
Zanesville 2	"
Columbus 2	"
Dayton 2	"
Cincinnati 2	"
Madison 3	"
Louisville 3	"
Nashville 4	"
Memphis 4	"
Jackson 5	"
Vicksburg 6	"
New Orleans 6	"
St. Louis 5	"
Harrisburg 1	"
Lancaster 1	"

Other dispatches between other places

	Per Word.
200 miles or under 1	cent.
500 or over 300 2	cents.
700 " " 500 3	"

1,000	"	" 700	4	"
1,500	"	" 1,000	5	"
2,000	"	" 1,500	6	"

10. It is used to a great extent, in conveying secrets of rise and fall of markets; for instance, a man may be purchasing goods in New York, gives his reference to the merchant—said reference being perhaps 700 or 800 miles away from him—by the aid of the telegraph he can know the standing of his customer, even before the purchase is completed. There are bankers, brokers, &c., in Wall-street, that receive and send, on an average, six to ten messages per day, throughout the year.

11. Yes, to a great extent. It oftentimes occurs that a party desires to "converse" with another 300 or 500 miles off. An hour is appointed to meet in the respective offices, and they converse through the operator. I have known instances of steamboats being sold over the wires—the one party being in Pittsburgh, the other in Cincinnati. Each party wrote down what they had to say, higgled awhile, and finally concluded the sale. Their correspondence was filed away, like other messages, and kept for reference, if ever called in question. It is often used by parties, when from home, corresponding with their families. Sometimes it is the messenger of a woe, deep and afflicting; and anon, that of heartfelt pleasure. In the early part of this year, the Astor House of New York, and the Burnet House of Cincinnati, had a series of telegraphic parties. An account of one of them, taken from the Cincinnati Gazette, is herewith appended—the parties conversing being about 750 miles apart —about twice the length of Ireland.

A Telegraphic Party.

On Friday evening of last week, by previous arrangement, the Astor House, in New York, and the Burnet House, in this city, had a family telegraphic party. They assembled respectively in the office of the O'Reilly line in each city.

They talked over family matters, interchanged congratulations, and drank each other's health, spending an hour in electrical and spiritual intercourse. As the spiritual was about commencing, the operators of the electrical in Philadelphia and Pittsburgh intimated their willingness to join therein; and forthwith orders were sent to the Monongahela House, at Pittsburgh, and Jones' Hotel, in Philadelphia, to furnish the needful champagne. It was done. Cincinnati then asks—"Are you all ready?" Pittsburgh answers—"Glasses charged." Philadelphia—"Corks just popping," and New York responds—"Aye, aye."

The following toasts were then drunk:—"'The O'Reilly Telegraphic Operators—Our thanks for exchanging family congratulations between the cities of Gotham and Porkopolis." To this toast the operators made appropriate responses, thanking the Astor and Burnet Houses for the champagne, hoping that all the "pain" of the respective families might be all "sham-pain"— and concluding with the following toast: "The Burnet and Astor Houses—May the skin of one gooseberry make "night-caps" for all their enemies." Drunk with three cheers. To this sentiment the two "Houses" replied simultaneously, and so highly complimentary to the operators, that they withheld the report, regarding it as "private and confidential." Here ended the first telegraphic party.

12. Where two sharp competing lines are in one city, strenuous efforts are made by both to be "ahead," especially on the arrival of a steamer bearing important commercial or political news. One illustration will suffice. Some time back the *Asia* arrived at Quarantine about 8. P.M.—was detained an hour by the health officer. The agent of the New York Associated Press and of the New Orleans Merchants' Exchange, Mr. Jones, to gain but a few minutes, had a boat in readiness when the *Asia* brought to. A small bag containing the latest news was handed over the steamer's side, to the small boat. By great exertions she gained New York half an hour ahead of the *Asia*. The bag was opened—a copy of her news was handed to us, addressed to the Merchants' Exchange, New Orleans, signed Jones—to work we went. It was being transmitted over the wires amid the thundering of the *Asia's* cannon, as she rounded the point; and a complete synopsis of her commercial and political news was received in Louisville, 1,100 miles in the interior, before the ship had actually reached the city. I may add that the telegraph would be more extensively used than even now by the

mercantile community, if its correctness and accuracy were improved. These inaccuracies arise altogether from the carelessness of operators, and not from any defect in the medium employed. These annoying drawbacks would nearly all, if not wholly disappear, by the enforcement of a more rigid discipline, and imposing a greater responsibility on the operators. Whichever line takes the lead in this much called for reform, will meet its due reward from the public.

13. Interruptions from atmospheric electricity have been greatly reduced of late, and it is confidently expected that they will, at no distant day, be entirely overcome. One of the means used at present is, by putting on a lightning arrester near to the recording instrument. This arrester is formed of a little glass globe, surrounded by a semicircle of small points, like needles, that approach quite close to the ball. These points carry off the lightning to the earth before it reaches the instrument, causing but a momentary interruption. The diminution of interruptions by thunderstorms has been reduced thirty per cent, by this means alone. Other causes of interruption are accidental,—trees, &c., falling on the wire and snapping it.

15. If a line is well insulated, it will work 500 miles on a single circuit. There are two lines in New York that work to Buffalo, the distance about 500 miles. Some lines that are in bad condition work but 300 miles, and others, with difficulty, 260 miles.

16. Yes; quite practicable, and coming more into use every day. One eighth of the dispatches between New Orleans and New York are in cipher. For instance, merchants in either city agree upon a cipher; and if the New Orleans correspondent wishes to inform his New York friend of the prices and prospects of the cotton market, instead of saying, "Cotton eight quarter—don't sell," he may use the following:—Shepherd—rum—kiss—flash—dog. The one message will come as correct as the other, and be wrapped in mystery to all through whose hands it may pass, and be only intelligible to those it is intended for. I have many times seen messages written in German and French sent over the wires, and it was "Greek" to all in the office. The

operators are not very fond of those kinds of messages; they prefer the English. During the business season at New Orleans, a great many messages, written in cipher, are received in this city on the eve of a steamship starting for England. They are mailed, and their answers may be brought back, by the returning steamship to Halifax, to New Orleans, couched in the same mysterious language. They traverse the wires from Halifax to New Orleans, impervious to the eye of the most curious.

ANSWERS FROM BAIN'S TELEGRAPH OFFICE,
29 WALL STREET, NEW YORK.

1. In answer to your first question, I would say that the greatest distance that messages are transmitted by telegraph, without rewriting, is on the lines between New York and Buffalo, a distance of about five hundred miles, and the Morse and Bain lines on this route transmit dispatches daily, without a repetition, and on a single circuit, without a repeating magnet. Several lines have worked a longer distance by the use of repeating magnets, as they are termed, which is done by making an electromagnet in one circuit break and close the next circuit, making the same vibrations on every magnet in the second circuit, as made by the operator in the first circuit. Under this arrangement, any number of circuits may be added on, until the lines reached round the globe, and the writing in the last circuit would be as perfect as in the first. Messages have been transmitted in this way over a thousand miles, without a repetition.
2. The tariffs of transmission vary materially in different sections of the country. It would be difficult to arrive at an average per mile throughout the United States. Every line makes ten words its maximum for one price; and the tariff from New York to Boston is twenty cents for the first ten words, and two cents for each additional word, a distance of 240 miles,† while the tariff from New York to Washington, 280 miles, is fifty and five cents. Competition has done much to reduce the rates on every route, except between New York and Washington, which

---

† Since reduced to fifteen cents for each ten words, and two cents for each additional word.

remains the same as when the first line started. The tariff from New York to Buffalo, 500 miles, is forty and three cents.
3. I believe the rules and regulations of every company give precedence only to the press and police, or official dispatches.
4. The difference as to rapidity between the instruments in use, is but little, if any, taking a day through. The House instrument, I think, is capable of transmitting more words in a minute, than either of the others, but to balance it a great deal of time is consumed in adjusting the instrument, as the instruments receiving and sending are required to run alike. From seventy-five to one hundred letters per minute, is probably about the number transmitted in ordinary business messages, on an average. The "fast method," as it is termed, invented by Mr. Bain, is capable of transmitting correctly 1,000 letters per minute, but the process of preparing the message to be transmitted, takes quite as long as to transmit it by either of the instruments.
5. Dispatches are transferred to writing as fast as they are received. They are either copied by the operator receiving, or read by him to a copyist as fast as received.
6, 7, 8, and 9, I suppose, refer entirely to the *New York Herald*, and which you can answer more correctly than I.
10. The telegraph is used by commercial men to almost as great an extent as the mail. This can be better illustrated by the number of messages sent and received between cities, where close commercial interests exist, during the hours between 10 A.M. and 5 P.M. For instance:—There are transmitted daily, between the cities of New York and Boston, between 500 and 600 messages, two-thirds if not three-fourths of which are transmitted between the hours above named. I know of some houses which pay from $60 to $80 per month to the telegraph: and I do not know but there are others who pay more. The amount paid by a commercial house is governed by the excitement there is in the market, of the particular article they may be dealing in. If there are "ups and downs " in the market, money is lavished upon the telegraph freely.
11. Many dispatches, of an entirely social character, are daily transmitted over every line; and, since the reduction on many

of the routes, they form a very large portion of the business. It is extensively used by the travelling public, who dispatch to their friends their progress, &c.

12. There are competing lines now in operation on every important route in the United States, except to New Orleans by way of Charleston and Savannah. The effect, on some routes, has been to double, if not treble, the business. In 1849, when there was no competing line between New York and Boston, the Morse line transmitted, on an average, between two and three hundred messages daily, while the average now transmitted on the three lines, is between five hundred and six hundred. The same ratio may be applied to all the other routes where the tariff has been lessened by the competition.
13. and 14. Interruptions of a few hours' duration are quite frequent during the hottest portion of the summer season from atmospheric electricity; but less so since the inventions of several ingenious protectors have been put in use. These protectors are made so as to bring several sharp points of metal connected with the ground as near as possible to a plate put in circuit with the wires, so that when the wires are overcharged with atmospheric electricity, it jumps off on these sharp points and passes to the ground. Interruptions most frequently occur from breaks in the wires, caused by hail storms, or the prostration of poles, wires and all, by the falling of trees, or the like.
15. I do not understand, from this question, whether you wish to know how many times communications are repeated between distant points, or how many times lines are interrupted. The latter I could not answer; and the former I am unable to state positively, but believe that dispatches are repeated some four or five times between here and New Orleans; there are four repetitions between this city and Halifax, and four between here and St. Louis.
16. Your sixteenth and last question I answer, by saying that it is practicable for correspondents to keep the subject of their dispatches from the operators and clerks—this is attained by the use of ciphers, which are quite extensively used between brokers and the like, both for the purpose of concealment and

to save expense, as they substitute a word for a sentence. I believe this is principally done for the latter, as persons using the telegraph must care less about having their business exposed to operators and clerks, or rather are better satisfied of the fact that telegraph operators and clerks, from their constant handling and reading of messages, become as it were dormant to the subject mentioned in the dispatch, they having no interest in the matter. I cannot illustrate this fact; but I know it to be one from experience. During the past four years I have had charge of one or more lines, and never as yet heard of the contents of a dispatch being divulged.

In great haste,   J. MC KINNEY.

* We understand that the Bain line between New York and Boston on the 17th of April, 1852, actually transmitted 500 messages, besides 5,000 words of foreign news, for the Associated Press.

ANSWERS FROM HOUSE'S PRINTING TELEGRAPH.

1. The longest line we have is about 1,000 miles—extending from New York to Cincinnati. Messages are transmitted that distance with ease; and no doubt we shall be able to telegraph direct from New York to St. Louis as soon as our line, now building, is completed.
2. The tariff is published complete.—(*See last part of this work.*)
3. 1st. Government (free). 2nd. Police. 3rd. Death. 4th. Press. All others, in the order received.
4. Thirty to thirty-five words, when written in full. A system of abbreviations used in news messages—say fifty words per minute. The proceedings of the Democratic State Convention, in the fall of 1850, containing 7,000 words, was transmitted from Syracuse to Buffalo in two hours and ten minutes; which was at the rate of about fifty-four words per minute. There is one rapid operator upon our Buffalo line, that has written 365 letters in one minute. There is a very marked difference in the speed of the different telegraphs. We prefer however not to state our opinion, fearing those who are interested in other systems would complain. We have our own views.
5. Not applicable to this line; the slips being delivered as printed

by telegraph.
7. On arrival of steamers; any sudden rise or fall in any leading article of merchandise.
10. The receipts from twenty leading commercial houses doing business through us, average $500 each per annum. We have a few customers whose telegraphing will amount to $1000 each per annum.
11. It is, to quite an extent.
12. There is competition on nearly all the routes. Its tendency to reduce rates has been equalised by an increase of business generally. Between New York and Buffalo the receipts of the Morse line in 1850, when there was no competition, were about $49,000. There are now three lines upon the same route, and the receipts of the united lines in 1851 considerably exceeded $100,000. Competing lines to a reasonable extent increase telegraphing, as business men rely with more confidence upon having their messages transmitted. Two lines upon the same route are rarely out of order at the same time; and as the lines have been made reliable, telegraphing has rapidly increased. Competition between New York and Boston has reduced prices altogether too low. The three lines have nearly all the business they can do, and sometimes more than they can do well; yet the receipts of each line are less remunerative than those of some other lines which do less work.
13. A heavy thunderstorm in the immediate vicinity of a line causes interruption, by charging the wires with electricity, destroying the effect produced by that from our batteries.
14. Other interruptions occur from falling trees, moving buildings, and carelessness on the part of the people in the country towns through which our lines pass.
15. Our system requires a complete circuit—way stations forming a part of, not breaking the main circuit. Copies can be dropped through the way offices.
16. It is practicable, though not of frequent occurrence. Correspondents often receive their own dispatches; the contents being known only to the operator writing them at a distant station. Half an hour's practice would enable a person to write his own

dispatches.

* By the slightest inspection of the wood-cut, representing House's Printing Telegraph, it will be seen that it has keys arranged in front like those of a piano, with the letters of the alphabet marked on them. An operator, in sending a message, merely touches or depresses them with his fingers like a person playing a tune on a piano, causing the letters to be printed by a similar instrument at the opposite end of the wire.

* The following are the press charges between the points named:
On the reports of Congressional proceedings and other news from Washington to New York, five cents per word up to the first 500 words; and one-third less for all words over 500 and under 1,000, and two-thirds reduction for all words over 1,000 and up to any quantity above that number. Between Philadelphia and New York, the House line charges one cent a word for the press.

* The Morse Washington and New Orleans Seaboard Line whose office is at the corner of Hanover-street and Exchange Place, make no deductions in favour of the press, long or short, and have but one list of prices for all messages.

* Each of the three lines—Morse's, House's, and Bain's—from New York to Boston, charge the press one cent per word on all news messages, of whatsoever length.

* The lines connecting Boston with Halifax, charge the Associated Press at a fixed rate per steamer, for all messages not exceeding 3,000 words.

* The Canadian lines make no distinction in their charges for tolls between private messages and those for the press.

Answers to Telegraph Questions made by Morse's New York and Buffalo Line.

1. Messages by telegraph have been actually transmitted over one continued circuit through 1,500 miles of wire. They are sent daily, and in the ordinary business of this line, 540 miles in a single circuit, without any re-writing.
2. Tariffs of transmission have no certain standard for computation. Competition reduces them materially. The average estimate may be as one cent (halfpenny) for ten miles on local business, and seven-tenths of a cent for ten miles for through business. The tariffs annexed to the list answer the inquiries. (*See prices at the end of this work.*)

3. Reports for the press, communications relative to sickness and death, and police messages, are entitled to precedence.
4. When intelligence is abbreviated, reports are sent at the rate of 2,000 words per hour; when not abbreviated, about ten words per minute. I have sent forty messages in forty consecutive minutes.
5. Morse's or Bain's telegraph cipher can be and are reduced to ordinary writing at the instant they are received—a copyist writing down words dictated to him from the slip of paper by the operator; and it requires a skilful and rapid penman to keep pace with the telegraphic transmission and reading off by the receiving operator.
6. Three thousand words of public business matter, and three thousand of reported, and about one thousand words relative to line business, in the daily adjustment of accounts, may be considered a fair day's work.
7. During the sittings of conventions, or elections, or the arrival of steamers, often from 2,000 to 8,000 words are reported. On some occasions of market excitement, the private messages are nearly doubled.
8. Debates of Congress are received at an average of about 4,500 words per day, and transmitted at the rate of 1,600 words per hour.
9. Two conventions exist between the leading New York press (seven morning papers now forming one, and two or three evening papers the other), who employ correspondents at important distant points to collect and forward the news; and other local agents to receive and re-write on manifold paper a copy for each paper in the association, distribute the same, and re-send the same news to the press at other points.
10. The telegraph correspondence of two or three private houses we could mention, amounts to about $1,200 per annum.
12. Between New York and Boston, and New York and Buffalo, and south to Philadelphia, there are three competing companies —and six wires for each route. The tariff from New York to Boston has been reduced from 50 to 20 cents per ten words for private messages. This is the greatest reduction made by

The Electric Telegraph in the United States

competition.
13. During the summer, interruptions occur about twice a week by atmospheric electricity. Sometimes the irregularity of current thus caused is entirely overcome by the adjusting skill of the operator. Sometimes, during thunderstorms, it is dangerous to attend to the instrument. Interruptions from the falling of trees, wearing out of poles, or the effect of violent storms, occasionally occur.
15. Eight or nine breaks occur between New Orleans and Quebec, when messages are re-written. In good weather, only four or five interruptions occur between the same places.
16. The concealment of the subject of dispatches is practicable; and although (unless sent in the cipher of a correspondent) they are necessarily known to the operators and copyist of the companies, yet instances have been rare in which publicity has been given to them.

13
SPECIMENS OF TELEGRAPH SIGNAL WRITING AND PRINTING

The following specimens of writing exhibit the alphabets of Swaim and Steinheil, and embrace an extension of the principles involved in each (as seen at p. 56 of this work), so as to denote all the letters required in the sentence, composed in their respective ciphers.

The reader will perceive, at the slightest glance, that the same signals, or ciphers for letters, can be changed at will, to represent other letters than those they stand for; and both Morse and Bain have occasionally modified their signal alphabets, an operation applicable to all kinds of cipher writing.

The Communication

INDEPENDENCE WAS DECLARED BY THE PEOPLE OF THE UNITED STATES ON THE FOURTH OF JULY 1776.

As Written by Swaim's Alphabetical Ciphers.

```
   I    n    d      e    p    e    n    d      e    n    c    e
  .|  ..|  ...|   |  ..|.  |  ..|  ...|   |  ..|  ...|   |   |
   w    a    s      d    e    c      l    a    r      e    d
 ...|  . |  ...   ...|  .|  |.  |  |..  ..|  |..   |  ....

   b    i    t      h    e      p    c    o      p
  ||... — ..|  |.  |||    |    ||.  |  ..|  ||  |.
   l    e    o      f      t    h    e    U
  ||..  |   ..|  |||..  ..|.  |||  |   ..|  |..
   n    i    t    e    d    S      t    a    t    e    s
  ..|  . |  ..|  |.  |   ....   |....|  |....|  |.  |  |...
```

(Cipher continued on page 122)

The Electric Telegraph in the United States

(Cipher continued from page 121)

```
  o      n       t       h    e       f     o           u
 ..|    ..|    ..||.   |||   |       |||.  ..|         ..||..
  r       t       h    J     u         l      i       s     e
 .||....| |.  |||  .|  ..|  |..      |...|   |...     |
  v     e     n        t    e     e    n         h        u
 .|||.  |   ..|      ..||.  |     |   ..|       |||     ..||..
  n     d       r      e     d    a    n         d     s    e    v
 ..|  .....||..      | .... ...|       ....    |...   |  .|||.
  e    n     t     y      s    i        x
 |    ..|  ..||   ..|    |....|        ||||
```

As Written by Steinheil's Alphabetical Dots or Signals, in two rows varying in perpendicularity as well as in numbers.

Independence was declared bi t
he people of the United States on
the fourth Juli seventeen hundred
and seventy six

As Written According to Morse's Alphabet.

I	n	d	e	p	e	n	d	e	n	c	e
w	a	s		d	e	c	l	a	r	e	d
b	y		t	h	e		p	e	o	p	l
e		o	f		t	h	e		U	n	i	t	e	d
S	t	a	t	e	s		o	n		t	h	e	4
t	h		o	f		J	u	l	y		1
7		7		6

Specimens of Telegraph Signal Writing and Printing

As Written by Davy's and Bain's Alphabets.

I	n	d	e	p	e	n	d	e
n	c	e	w	a	s	d	e	c	l
a	r	e	d	b	y	t	h	e
p	e	o	p	l	e	o	f	t	h
e	U	n	i	t	e	d	S	t	a
t	e	s	o	n	t	h	e	4
t	h	o	f	J	u	l	y
1	7	7	6

As Printed by House's Electrical Printing Telegraph.

INDEPENDENCE WAS DECLARED BY THE PEOPLE OF THE UNITED STATES ON THE FOURTH OF JULY SEVENTEEN HUNDRED AND SEVENTY SIX

14

CONNECTION OF THE PRESS WITH
THE ELECTRIC TELEGRAPH—EARLIEST COMMUNICATIONS BETWEEN NEW YORK AND OTHER POINTS—CROSSING THE NORTH RIVER SYSTEM OF NEWS REPORTING CIPHERS, ANECDOTES AND INCIDENTS

The first telegraph office ever opened in New York, was in the basement of No. 10 Wall Street. This was selected for the reception of messages by the Morse New York and Washington line. The operating office was in Jersey City, where it remained for a year or two before any successful attempt was made to cross the North River, which at all times has caused much annoyance to the Southern lines. The Morse company, to enable them to establish a permanent communication over the river, extended a line from New York up the eastern shore of the Hudson for near 60 miles, when reaching a narrow part of the stream, with great elevation of land on either side, they stretched a wire across it, and from thence continued their line chiefly on the western bank of the river, southward, until it formed a junction with their main line terminating at Jersey City. They had previously experimented with imperfectly insulated wires sunk on the bottom of the river, but they were either hooked up by the anchors of vessels, or failed properly to conduct the fluid. The long extension of their line up and down the river also failed to work satisfactorily, and to keep it in repair and working order was found very expensive. Some party finally suggested the use of Gutta Percha as a coating for wires, and to be used for submarine purposes. It was tried and found to answer remarkably well. And wires coated with this material are still used as means of communication from the Jersey shore to New York, anchored on the bottom of the Hudson, and in places twenty or more feet below the surface. Occasionally one is drawn up on an anchor, but by having two or three wires immersed at a distance from each other, if one is injured, resort can be had to the others. This discovery of the non-conducting properties of Gutta Percha was an important step in the progress of telegraph lines, and all the conducting wires about telegraph offices are coated with this material. Its first application was of American origin, it was afterwards adopted in England, and was successfully used in combination in insulating the wires stretched across the channel between France and England.

We have used the same substance for insulating lightning rods attached to our dwelling with success. A stout wire or small rod may be coated its whole length, or be made to pass through staples coated with it. The coated wire may have a free sharp point, secured above the chimney top, and afterwards carried through in any direction to reach the ground, or nailed by strips to the house, while the lower end being free, is sunk a proper depth in the ground. This new plan will be found cheaper and better than any other.

House's New York and Philadelphia Telegraph Company crossed the Hudson near Fort Lee, about ten miles above the city, by erecting high masts on either side, and stretching a steel wire from one side to the other; allowing for its sag, the length being about a mile. The most frequent interruption occurred in winter, when, from the accumulation of sleet or ice on the steel wire, it would break. The Bain Company also crossed the river near the same place, by a somewhat similar method. Owing however to frequent interruptions to the wires extending over the river in all forms, all three of the lines found it necessary to keep offices and instruments in Jersey City, to which they could repair when necessary, and this became quite frequent in winter.

It was early in the autumn of 1846, when the writer of this handed in his first message for the newspaper press at 10 Wall Street. It contained a brief account of the launch of the U. S. sloop of war *Albany* at the Navy Yard Brooklyn, and was directed to the *Washington Union*.

In 1847, the three Morse lines then in existence concentrated in Post's Buildings, corner of Hanover Street and Exchange Place. The following year they scattered, and the ten or twelve different lines now occupy offices at a number of places; but which are chiefly confined to Wall Street, and its vicinity.

Soon after the first lines of telegraph were put in operation, it became apparent that they would necessarily become important media for the transmission of news for the press.

At the commencement, the papers proceeded to employ them with some caution and hesitation; their dispatches were usually brief, and as much condensed as possible. The expenses were heavy, and only a few papers in each city at first employed the lines to any material extent. Among the earliest and most liberal patrons of the telegraph, were a few members of the New York press.

When the telegraph was set to work between New York and Philadelphia, and afterwards had progressed as far south as Richmond and Petersburg, in Virginia, the Mexican war was in full blast, and hence its utility to the press in forwarding army news was such as, in a measure, to force them into its employment.

The free use made of the telegraph, by the papers referred to, caused others to enter the field, and finally to unite with them in their news arrangements. They ultimately adopted a plan of running a daily horse express between Mobile and Montgomery, in Alabama (about two hundred miles), to expedite the news to the southern terminus of the telegraph line, in advance of the mail. This, however, when the line was extended sufficiently far south, was abandoned.

When the line was carried east to Portland, in Maine, the press then, which comprised all the chief morning papers of New York, ran an express from Halifax, on the arrival of each steamer, to Annapolis, from whence a steamer conveyed it to Portland, from thence it was conveyed by telegraph to Boston and New York. The leading Boston papers also participated in this arrangement. The cost of getting the news in this manner averaged near $1,000 per steamer. When the wires reached St. John's the expenses were reduced, and finally, when they reached Halifax, they were brought down to the simple cost of transmission, or to about $500 per steamer. Mr. D. H. Craig was the special agent of the Associated Press in these enterprises.

At the outset there was a want of system in the collection, transmission and distribution of telegraph news for the press. It became apparent that the employees in the telegraph offices could not be expected to collect news at important points, and forward it. Their occupation confined them to the immediate duties of their offices. Hence the business of telegraphing brought into requisition Telegraph Reporters. We were among the earliest to engage in the occupation; we commenced with the commencement of telegraphs, and when the whole system was new and imperfect, and in a manner without organisation. We early invented a kind of short-hand system, or cipher, intended greatly to abbreviate commercial news transmitted by telegraph, a notice of which appeared in the *Herald* in 1847. This was so arranged, that the receipts of produce and the sales and prices of all leading articles of breadstuff's, provisions, &c, could be sent from Buffalo and Albany daily, in twenty words, for both cities, which, when written out, would make one hundred or more words. This plan of abbreviation, or some modification of it is continued on the same route, besides others, to the present day. Copies of the cipher, either in manuscript or in print, were placed in the hands of correspondents who could either compose or translate messages for the press. Another party also contrived a cipher, but on an entirely different plan. Our first effort was found in its daily use to be imperfect, and we soon prepared a second edition. We commenced sending and receiving commercial reports by it early in 1847, between New York, Baltimore, Boston and Buffalo, and subsequently between New York and Cincinnati, New Orleans and St. Louis.

Mr. F. O. J. Smith, the controller of the New York and Boston Morse line, established his charges at fifty cents for each ten words. We received a daily report from Boston of the markets over his line, of ten words in our cipher, which, when translated for the press, made at least from fifty to sixty.

Mr. Smith thinking we were getting more than our money's worth, decided that five letters constituted the average of English words, and directed that all the letters in a message sent in cipher, should be counted, and the whole divided by five for the number of words, and charged accordingly. We then, thinking other lines might follow his example, sat down and ransacked Walker's Dictionary for a collection of short words, and in no case as far as practicable did we select one with more than five to six letters. After much labour we had a new cipher ready for the press. When printed it made about 70 or 80 pages octavo, and altogether, the edition comprising only a few hundred copies for private use, cost us a considerable sum of money, on account of the large amount of figure work. Mr. Smith soon after decided that three letters made an English word, and we then abandoned receiving markets over his line from Boston, but supplied some of the papers in that city for some time with New York markets prepared by it, and also used it on lines leading to Baltimore, Cincinnati, and Buffalo. We have also made another manuscript edition of our cipher, which we have still on hand.

In preparing ciphers, only plain English words could be introduced, and so arranged, that one could stand for, or be interpreted to mean half a dozen others. Thus taking flour, the first word beginning with a consonant, would express the condition of the market, the second also beginning with a consonant, but with a different signification, would indicate the price, and the third commencing with a vowel, would in all cases express the quantity.

The following may suffice as examples of ciphers for flour, wheat, and corn, with an abbreviated table for quantities. From these specimens it may readily be seen, that a similar compilation of words beginning with consonants, can be greatly extended, and made to embrace all the conditions and prices of every leading article of commerce, and the quantities affixed to words beginning with vowels, can be extended to much larger quantities. The samples annexed are very brief extracts from a complete book of ciphers, in which the words expressing the conditions of the market for each article, are largely multiplied; such as those beginning with *B* in alphabetical order, for the state of the flour market. The words commencing with *C* are also multiplied to express its prices. The words beginning with this letter are sufficiently numerous to give the prices for all the different kinds of flour, from the lowest to the highest grades. When the words beginning with *C* have given figures for a particular kind of flour, beyond which it is not likely to sell, a heading for another grade is introduced, such as "Michigan" or "Ohio," and other words beginning with *C* continued in alphabetical regularity, to express the prices appertaining to each. The same course is continued when we come to wheat, where words commencing with *K* stand for general commercial remarks, and those beginning with *D* for prices. In the case of corn, words commencing with *L* stand for general remarks regarding the trade, and those beginning with *F* for prices. In each case it is apparent, that the words can be so multiplied, as to embrace all general remarks applicable to the markets, as well as to prices and quantities, while quantities in all cases are taken from the table of words beginning with vowels; which table, also, can be extended so as to embrace the highest amount of figures necessary to be used. Proper words can also be employed to express the fractional parts of prices, as in cotton sterling exchange, &c. The foregoing conditions were carried out in our printed book of ciphers.

Examples from System of Commercial Ciphers.

FLOUR

The following words can be used as indicating the condition of the flour market.

Baal.	The transactions are smaller than yesterday.
Babble.	There is a good business doing.
Babe.	Markets dull; buyers do not enter freely at the higher rates demanded.
Baby.	Western is firm, with moderate demand for home trade and export.
Bare.	Firmer, with fair home demand, including lots for export.
Back.	Market shade firmer; but owing absence private advices, buyers and sellers do not meet.
Bacon.	Dull, but if anything shade firmer.
Bad.	Market for common and fair brands of Western is lower, with moderate demand for home trade and export.
Badge.	Fair demand for Western flour; market without material change.
Badly.	Market has been rather heavy, but we notice no material change in prices.
Basin.	There is a speculative demand at better prices.
Button.	Market quiet and prices easier.

Connection of the Press with the Electric Telegraph

Prices—PURE GENESEE, *in Albany.*

Cairn,	$4.37	Calx,	$5.06	Car,	$5.75
Cajole,	4.44	Came,	5.12	Carat,	5.81
Cake,	4.50	Camp,	5.18	Carbon,	5.87
Calf,	4.56	Can,	5.25	Card,	5.94
Calid,	4.62	Canal,	5.31	Care,	6.00
Calif,	4.69	Cane,	5.37	Cargo,	6.06
Calk,	4.75	Canoe,	5.44	Cark,	6.12
Cale,	5.81	Canon,	5.50	Carle,	6.18
Calm,	4.87	Cant,	5.56	Carol,	6.25
Calp,	4.94	Cap,	5.62	Carp,	6.31
Calve,	5.00	Capon,	5.69		

WHEAT

If it be necessary to give "half cent" quotation, add the termination of "ed," as "decayed," 126½.

Kale.	Dull, but prices are firm.
Khan.	Firm, with good milling inquiry.
Kata.	Supplies of Western mixed are larger, and prices heavy.
Kaw.	Steady and firm.
Keck.	Fair demand; firmness holders checks operations.
Kedge.	Fair milling demand for prime.
Keek.	Prime Ohio in good demand.
Keel.	Moderate inquiry; market steady.
Keen.	Prime in fair demand; market firm; common descriptions dull, with downward tendency.
Keep.	Firm, but dull.
Keg.	Held above the views buyers.
Kelk.	Not much inquiry.
Kell.	Demand only for prime parcels, which are scarce, and held above views buyers.
Kelp.	Wheat easier, especially for low grades.

GENESEE.

Dale,	97	Date,	109	Deal,	121
Dally,	98	Dater,	110	Dean,	122

Dam,	99	Datam,	111	Dear,	123
Dame,	100	Daub,	112	Debar,	124
Damp,	101	Dauber,	113	Debit,	125
Dance,	102	Dauby,	114	Decay,	126
Dank,	103	Daunt,	115	Decent,	127
Dare,	104	Davit,	116	Deck,	128
Dark,	105	Daw,	117	Decoy,	129
Darn,	106	Dawn,	118	Deem,	130
Dart,	107	Day,	119	Deep,	131
Dash,	108	Deaf,	120	Defer,	132

Prefixes

"re," ½ a 1.   "de," ¼ a ½.

Terminations

"ed," ½.   "able," 1.   "ing," 1½   "ment 2.

CORN

☞ See instructions as to Wheat.

Label.   Is steady at yesterday's rates.
Labor.   Good inquiry.
Lac.   Easier, with better inquiry.
Laced.   Held higher; transactions limited.
Lacing. In brisk demand.
Lack.   Firm, and in good demand.
Lade.   Very firm.
Laden.   Heavy, owing to large receipts.
Ladle.   Good request at better prices.
Lag.   Market lower.
Lagger. In fair request.
Laid.   Not so active, but without change to notice.
Lain.   Foreign news; unsettled market; no sales of importance made.

YELLOW CORN

Fang,	57	Fault,	67	Feign,	77
Farce,	58	Faun,	68	Feint,	78
Fare,	59	Favor,	69	Fell,	79

Connection of the Press with the Electric Telegraph

Farin,	60	Fay,	70	Felt,	80
Farm,	61	Feal,	71	Fen,	81
Fash,	62	Fear,	72	Fence,	82
Fast,	63	Feast,	73	Feud,	83
Fatal,	64	Fed,	74	Fenny,	84
Fate,	65	Feel,	75	Feral,	85
Fatly,	66	Feet,	76	Ferny,	86

Prefixes

"ex," delivery within a few days. "un," delivery during month. "in," delivery during next month. "re," ½ a 1. "de," ¼ a ½.

Terminations

"ed," ½   "able," 1.   "ing," 1½   "ment," 1 to 2.

The following may represent quantities of any article sold:

Abot,	100	Admit,	3,000	Agage,	10,000
Abaft,	200	Adduce,	3,500	Aid,	11,000
Abuse,	300	Ache,	4,000	Aim,	12,000
Abase,	400	Acute,	4,500	Ail,	13,000
Abash,	500	Adept,	5,000	Air,	14,000
Abate,	600	Adore,	5,500	Agast,	15,000
Abide,	700	Adom,	6,000	All,	16,000
Able,	800	Adult,	6,500	Alert,	17,000
Abode,	900	Affix,	7,000	Allow,	18,000
About,	1,000	Afloat,	7,500	Alone,	19,000
Abut,	1,500	Aft,	8,000	Apart,	20,000
Abond,	2,000	Age,	8,500	Alude,	21,000
Adapt,	2,500	Agape,	9,000	Amaze,	22,000

If we send the following words by telegraph, *"bad, came, aft, keen, dark, ache, lain, fault, adapt* (nine words,) the following will be the translation: *"Flour Market for common and fair brands of western is lower, with moderate demand for home trade and export. Sales 8,000 bbls. Genesee at $5 12. Wheat, prime in fair demand, market firm, common description dull, with a downward tendency, sales 4,000 bushels at $1 10. Corn, foreign news unsettled the market; no sales of importance made. The only sale made was 2,500 bushels at 67c."* (sixty-eight words). On some lines each figure used is counted as a word, while in other cases they are only counted for the number of words in which they can be spelt. In translating a cipher message many terms have to be understood, such as bbls., bushels, hhds., cask, &c.

When competition in telegraph companies sprang up, the rates of tolls in many directions became materially reduced. Hence between New York and Boston, ten words, which formerly cost fifty cents, or five cents a word, are now for private parties forwarded for twenty cents, or two cents per word, and all newspaper messages are sent for one cent per word. The sea-board line via Washington to New Orleans, make no deduction for the press, and charge $2.40 for ten words, and fourteen cents for each additional word on through messages. When the first line was opened to Washington, the charges were high and no deduction allowed on dispatches for the press; hence, it became necessary to form congressional ciphers; several parties attempted to arrange them; among others was Mr. Wills of Baltimore, who made a very good one. We without any knowledge (at the time) on our part of the plan upon which he had proceeded, went to work to make one for our own use, and printed it. In forming this cipher, one word, as in the commercial cipher, was made to express many others. A few samples taken at random from our printed book, will suffice to explain its nature.

Senate.

Babble.	From the committee on finances, reported.
Babe.	From the committee on commerce, reported.
Babel.	From the committee on manufactures, reported.
Bacon.	From the committee on agriculture, reported.
Bad.	From the committee on military affairs, reported.
Badge.	From the committee on naval affairs, reported.
Badly.	From the committee on militia affairs, reported.
Baffle.	From the committee on public lands, reported.
Beal.	Reported a bill authorising the Secretary of the Treasury.
Beak.	Reported a bill authorising the Post Master General.
Beam.	Reported a bill authorising the Secretary of the Navy.
Beard.	Reported a bill authorising the Secretary of War.

Connection of the Press with the Electric Telegraph

Beat.	Reported a bill authorising the Attorney General to —
Bed.	Reported a bill authorising the Secretary of State to —
Beer.	Reported a bill authorising the President to —
Basket.	The resolution from the House was taken up.
Baste.	The Chair laid before the Senate a communication from —
Bath.	The debate was continued to the hour of adjournment.
Bating.	The Senate asked for a committee of conference on —
Battle.	The Senate agreed to a House proposition for a committee of conference on —
Bawd.	Presented resolutions adopted by the Legislature of —
Bawl.	An interesting debate followed, in which several Senators participated.

Resolutions.

Bent.	Submitted a resolution that the President inform the Senate.
Benumb.	Submitted a resolution that the Secretary of the Treasury inform the Senate.
Beray.	Submitted a resolution that the Secretary of the Navy inform the Senate.
Bereft.	Submitted a resolution that the Post-Master General in form the Senate.
Berth.	Submitted a resolution that the Secretary of War inform the Senate.
Besot.	Presented the credentials of —
Bestow.	Presented the credentials of his colleague.

Appropriations, &c.

Braze.	The naval appropriation bill was taken up, and —
Bread.	The Indian appropriation bill was taken up, and —
Brew.	The army appropriation bill was taken up, and —
Bribe.	The bill to reduce and graduate the price of public land was taken up, and —
Brick.	A bill in relation to the Military Academy at West Point was taken up, and —
Bride.	The river and harbour bill was then taken up, and —

Motions.

Cinder. Motion referred to the committee on foreign relations.
Cimeter. Motion referred to the committee on patent office and post-roads.
Civil. Motion referred to the committee on claims.
Clad. Motion referred to the committee on pensions.
Claim. Motion referred to the committee on military affairs.

Miscellaneous.

Cave. The resolution referring the President's message to appropriate committees, was then called up.
Cavern. The Speaker called him to order.
Cavil. The bill granting bounty lands to soldiers, was then taken up.
Cellar. Resolutions came up for consideration.
Cent. Several messages in writing were received from the President of the United States by the hands of his private secretary.
Chaff. Gave notice that he would at an early day ask leave to introduce a bill.
Chain. Then addressed the committee.
Chant. T he Journal was then read and approved.

In transmitting reports, the blanks intervening between the cipher word and the explanation, was filled with the actual name of the member connected with the subject, giving his surname only, unless either house contained more than one party of the same name, and then the first name was given.

The whole of the cipher words were arranged in alphabetical order, both for the Senate and House, and made to embrace almost every variety of proceedings, with parliamentary, judicial, diplomatic, and executive phrases, and in some cases the heads of debates pro and con, in reference to well-known party subjects. We prepared a cipher for legislative proceedings, assisted by Wm. Lacy, Esq., of Albany. We made two manuscript copies, which are still preserved. The proceedings of the long sessions of the New York legislature of 1849 and 1850, were transmitted by it; but the reduction of

Connection of the Press with the Electric Telegraph

tolls caused by the competition of the Bain and House lines, rendered their further use unnecessary.

The press at first, owing to the expense, would not agree to receive more than would make from the half to one column of the *Sun* newspaper. We then supplied them under a weekly contract, and paid our own tolls and reporters' fees, in all directions. When we received a scrap of news we endeavoured to make the most of it the facts would justify, by writing it out as full as possible. Thus often, from a small page in manuscript of congressional reports in cipher, have we written out enough to fill a column of the *Sun*. On one occasion we fell into an unpleasant error. A certain quasi-democratic measure was taken up, and just as it was announced that Reverdy Johnson (a Whig of Maryland) had risen to speak to it, the wires gave out. On looking over the heads of our ciphers to find out if possible what a Whig would likely say in opposition to it, we briefly endeavoured to make the honourable member say some very clever things against it, which duly appeared next day in the city papers which we were supplying. When the mail came to hand we were chagrined to find, that Mr. Johnson so far from making a speech against the measure, had actually advocated it. We also received a letter from our Washington reporter, begging us never again to take any matter for granted; otherwise he should lose his credit, and be ruined in his business. We never after that attempted a similar liberty. As good luck would have it the affair blew over without exciting any public remark.

The erection of competing lines, with an increase of wires on the first, and introduction of better modes of insulating them, led to such a reduction of tolls on reports for the press, that our congressional ciphers were laid aside. The reports were afterwards obtained as we now see them.

It was several months after the first New York and Washington line was opened, before scarcely any reliance could be placed on it. And the few brief messages we sent to the Union were suspended, on the ground of their being received in an unintelligible form. Those received in return by us from the south for the New York press, were equally blind and puzzling. We recollect on one occasion sitting up to a late hour at night, waiting for a piece of news. When, however, it came to hand, we studied over its meaning till our head ached. We finally, in a fit of despair, made as near as we could a *facsimile copy* of it, and sent it to the press. It was not long before a printer's devil from one of the offices returned with the copy, and said, "The foreman, sir, says that he cannot read your copy." "Please give him our compliments," we replied, "and tell him that we cannot read it ourselves."

This irregularity and illegibility was owing to the imperfect construction of the early lines, and to the want of experience on the part of operators, copyists, &c.

In copies of our congressional cipher, the word "Dead" occurred, opposite to which stood the following explanation: "He," (Blank, a member of the Senate), "after some days absence from indisposition, reappeared in his seat." The Hon. John Davis, a senator from Massachusetts, had been confined by illness, but became convalescent, and reappeared in his seat. Hence the reporter in Washington telegraphed, in his report of the Senate proceedings, "John Davis Dead," which, in the hurry of writing out, was sent to the press literally as it came. The same message was overheard in the telegraph office in Philadelphia, by another reporter, who also gave a literal copy to the papers there. A literal copy also reached Boston. It drew forth a great number of the most complimentary eulogiums on the supposed deceased from the press, which Mr. Davis had the satisfaction of reading. The error, the following day, was corrected by telegraph as far as practicable. Mr. Davis is still alive, and a member of Congress (1852).

In June 1848, the Whig Convention met in Philadelphia, and it became extremely doubtful who would receive the nomination for the presidency. It seemed to lie between Mr. Clay, Gen. Taylor, Gen. Scott, and Judge McLean. The day on which it was believed a decision would be made, we devised a plan to get the news of the event in advance of all others. There was then no line across the Hudson River; and when the news arrived in Jersey City, it would have to be sent over on a ferry boat. We therefore provided a set of flags. That mounted with a single piece of white cloth, was to indicate that Gen. Taylor was nominated. Another, with a red flag, was to denote that Mr. Clay had received the nomination.

Two white flags on the same staff was to indicate Gen. Scott as the nominee, and two red flags on a staff were to denote the result in favour of Judge McLean. We took a young man from the Courier Office, gave him a white signal flag, to answer us when we should take our position on the west side of the River. We placed him on a pier near the Courtland Street Ferry, and gave him the necessary instructions. About 10 to 11, A.M., we crossed over. It seems that there was, unknown to us, a party employed by a company of brokers, to telegraph the price of stocks from the top of the Merchants' Exchange to Jersey City. One man would stand on the top of this building, and, on waving his flag, would be answered by another man, who would wave a white flag as a token of his readiness to receive communications. The figures of prices of stocks would be indicated by the motions given to the flag from right to left, horizontally, &c., by the operator. The man who received the numbers in Jersey City, would re-send or telegraph them to Philadelphia. After I had placed my young man, at about 11½, A. M, the brokers' man in Jersey City stepped out on the pier, and waved his flag to the man on the Exchange as usual, as a token that he was on hand. Our young man, seeing this white flag waving on the pier, supposed it to be ours, and immediately ran to the newspaper offices, and informed them that Gen. Taylor was nominated. It produced great excitement, and was telegraphed east. On reaching Portland, Maine, 100 guns were fired. The eastern line soon after gave out, and the news could not be contradicted. It turned out that Gen. Taylor was not nominated until the succeeding day.

We could relate many other curious incidents in connection with early telegraph operations, many of which are quite amusing.

We afterwards ascertained that a combination of brokers, in Wall Street and Philadelphia, had previous to the erection of the electric telegraph line between New York and the latter city, established a private semaphoric or *visual telegraph*.

Their plan was to station men on eminences at every six or eight miles, with telescopes in their hands. They were also provided with flags. At the first call of stocks at 11 o'clock in the Board of Brokers, at the Merchants' Exchange, the prices at which the stocks chiefly dealt in, sold at, would be given to the flag-man on top of the Exchange, who would communicate the quotations to another flag-man in Jersey City, or on top of Bergen Hill, and he to the next, until Philadelphia was reached. The sales in Philadelphia would reach New York by a similar method. The order in which they were sent, indicated what stocks were referred to.

The flag was attached to the end of a short stick, held in either hand. All the figures could be represented by the position in which the flag was held or moved. Thus, if the flag was held at right-angles to the body, it might be taken to mean any one of the ten digits. When held up directly over the head, it might indicate a second figure. Other figures could be represented by changing the flag to the left hand, and varying the motions accordingly. When held perpendicularly over the head, it might mean that the market generally was up; and when held down at a sharp angle to the body, it might indicate the reverse. Half revolutions to the right, would give half numbers; the same to the left, quarters—and over the head, *eighths*.

It was said that the news by this plan could be transmitted in about thirty minutes. The chief originator and superintendent of the scheme was said to be a broker in Philadelphia, who retired with a fortune. Lottery dealers, also, contributed to sustain that or a similar line, for the purpose of transmitting the drawn numbers of successful tickets. Some wealthy men in Jersey City, at one time extensively engaged in lotteries, it is said, derived important advantages from this kind of *visual telegraph*.

Within a year or two after we had engaged in telegraph reporting, an association of three or four reporters was organised, who employed others in various important localities to forward and receive news for the press. Their services were either paid for at stated weekly salaries, or in steamers' and other news forwarded in exchange. The press were charged so much per week each, while the reporters paid tolls and all other expenses. This association only lasted about twelve months, when it was dissolved, and, as far as New York was concerned, we became the agent of the New York Associated Press, for all news arrangements of a commercial and miscellaneous character throughout the United States. A committee of the Press attended to the foreign news received by steamers at Halifax, which was the more necessary, owing to difficulties which had arisen between them and Mr. F. O. J. Smith, the president of the Morse Eastern Lines.

The Association thus formed consisted of six morning papers, viz: *Herald, Sun, Journal of Commerce, Express, Tribune,* and *Courier* and *Enquirer*. The *Times* has since joined the Association, making it consist of seven instead of six.

We received and distributed the news, paid all tolls and other expenses necessary to conduct the business. We employed reporters in all the principal cities in the United States and Canada, and, on receiving it in New York, would make about eight or nine copies of it, on manifold paper—six for the New York press, and the remaining copies for re-forwarding to the press in other cities and towns. To this had daily to be added the New York local and commercial news, ship news, &c. The remuneration for services was made to depend chiefly upon what we could obtain from papers in other cities, such as Boston, &c, for the news of all kinds re-forwarded, including the local intelligence. The agent had an office separate from the press, but centrally located, where he employed generally an assistant, besides one or two other parties either as clerks or aids, with an errand boy or two.

In reporting Congressional proceedings, the usual plan was to employ two reporters in Washington; one for the House of Representatives, and another for the Senate. The reports of the House would be sent by one line, and those of the Senate by the other.

The plan upon which the members of the Associated Press acted in obtaining extra news, was such, that any one of them could order any particular kind of news, such as proceedings of conventions, &c, and the others were to exercise their option whether they would take it. If all should decline it but one or two, those one or two were expected to bear the whole expense.

One of the earliest telegraph feats, after the extension of the telegraph lines west to Cincinnati, was brought about by the agency of the *New York Herald*, and before any regular association of the press was formed in New York.

It became known that Mr. Clay would deliver a speech in

Lexington (Ky.), on the Mexican war, which was then exciting much public attention. Mr. Bennett, editor and proprietor of the *Herald,* requested us to have Mr. Clay's speech reported for the paper. We at once proceeded to make arrangements to carry it into effect. We had a regular and efficient reporter already employed in Cincinnati, a Mr. G. Bennett; we also had a Mr. Thompson in Philadelphia in cooperation with us for some papers there, and which agreed, if the speech was first received, to share the expense with the *Herald.* The *Tribune* in New York, and the *North American* in Philadelphia, agreed to start for a report of the speech, in opposition. From Lexington to Cincinnati was eighty miles, over which an express had to be run. Horses were placed at every ten miles by the Cincinnati agent. An expert rider was engaged, and a short-hand reporter or two stationed in Lexington. When they had prepared his speech, it was then dark. The express-man, on receiving it, proceeded with it for Cincinnati. The night was dark and rainy, yet he accomplished the trip in eight hours, over a rough, hilly country road. The whole speech was received at the *Herald* office at an early hour next morning, although the wires were interrupted for a short time in the night, near Pittsburgh, in consequence of the limb of a tree having fallen across them. An enterprising operator in the Pittsburgh office, finding communication suspended, procured a horse, and rode along the line amidst the darkness and rain, found the place, and the cause of the break, which he repaired: then returned to the office, and finished sending the speech.

The Philadelphia *North American*, upon whom the *Tribune* chiefly depended, failed to get its report; and the latter purchased a copy from the *Herald*.

The expense in securing the speech by express and telegraph, amounted to about $500.

The telegraphs have derived a very large share of their revenue from the press. The whole expense, for telegraph reports of all kinds, have some years cost the New York Associated Press (six in number) probably about $5,000 each, or a total of $30,000 per annum. The average for the past five years probably has not been less than about $25,000 to $30,000 per annum. During long sessions of Congress it exceeded this amount.

Sometimes a single paper availed itself of the privilege of ordering long and expensive reports of meetings, speeches, conventions, &c., in which its associates participated or declined as best suited their estimate of the value of the news. In case the other papers refused to receive it, the whole expense was borne by it. The *Herald* is the only one of its associates which publishes a Sunday paper—hence it takes all the telegraph news which is received on Saturday afternoon and night, and pays the whole expense of the tolls.

In managing the financial affairs of the business, we soon found it necessary to arrange a system for conducting it, and afterwards rigidly adhered to it.

We proceeded thus. All the telegraph offices in the city were required to send in their accounts weekly. These were examined and all set down in a general bill, to which was added office and other incidental expenses. The aggregate sum was then divided into six parts, and submitted to the treasurer of the Association, on whose approval the respective amounts were collected from the associated papers. On each Saturday we paid off all the telegraph bills, and office expenses, and commenced the following week *de novo*. In case any paper had failed to pay, its news would have been stopped. Papers supplied in Boston and various other places, were expected to pay monthly. At the expiration of each month their accounts were regularly forwarded. If any country paper failed to make monthly settlements their news was discontinued. The bills of our correspondents for services, were also paid monthly.

The system thus organised worked quite smoothly, and gave very little trouble: our services, however, were severe. Help, with the proper tact and necessary prior instruction, could not be had. We were compelled to take inexperienced and youthful parties, and by long drilling to create as it were our assistance. We had also to hunt up and secure the best Reporters in all parts of the country.

The business required our personal attention day and night, Sunday and Monday. Often on stormy nights in winter, when our errand boys were either ill, or absent in Jersey City, have we gone round at twelve and one o'clock, and delivered messages with a snow or sleet storm beating in our face; and having, at many of the offices, to climb three or four pair of stairs to find the composing room. For months at a time we seldom retired before twelve to one o'clock, and then had to be on duty through the next day. During state elections we were frequently up all night; and at the Presidential election in 1848, we remained up for three nights consecutively. The tolls on the returns on that occasion cost the associated press, something over $1,000.

Our services were thus continued, until the 19th May 1851, when we resigned the general news agency, after having devoted from five to six years of unremitted health-wearing toil to the business, and were succeeded by D. H. Craig Esq., who continues to act as such at the present time; though we continue to prepare the commercial news for the associated press, which is sent off daily through Mr. Craig in all directions as previously done by us. This change has relieved us of night work and many annoyances which seemed in a measure inseparable from the business.

The vexations endured were aggravated by dissensions which grew up between the managers of some of the Morse telegraph lines and the press. There were also contentions between the members of the press in Boston and other places, fanned if not engendered by the jealousies of some of the Morse lines, and especially by those under the control of F. 0. J. Smith. This gentleman refused to have steamers' news come over his line from Halifax, for the associated press, unless they dismissed Mr. Craig, then acting as their Halifax agent. This led to a rupture, by which the press of Boston became divided. The Association retained Mr. Craig, and ran a locomotive express at an enormous expense with each steamer's news, from Portland to Boston, there being no telegraph between the two points but that owned by Smith. From Boston it came over by the Bain's line to New York. The Association also, by its encouragement, caused a company to extend the Bain line from Boston to Portland, where it united with the lines extending thence to Halifax, and which were beyond the control of Smith. The war was a very fierce one; many pamphlets appeared on both sides, including one by Mr. Craig in his defence against Smith's charges. The latter left no stone unturned. Among other efforts to thwart the Association, it is said that he endeavoured to get control of one of the links on the Halifax line east of Portland. He also appealed to the Provincial Legislature of New Brunswick, and protested against the management of the Halifax line by its superintendent; but all without avail. His success in putting the newspaper press by the ears, was not only less difficult, but more complete. At one time Smith refused to receive and transmit private messages handed in by merchants and others for Halifax, or to let anything come over his line from thence. The Morse New York and Buffalo line was managed on a system almost as peculiar, under the direction of its president, T. S. Faxton. When the Bain and House lines were completed, between New York and Buffalo, and came in competition with his own, he demanded that unless all the news we were in the habit of sending for the papers west, went over his line, that he should decline sending our reports unless they were prepaid in advance at private rates!

Some of the press desired to get our news over the Bain or House lines. It was in vain we urged upon Mr. Faxton, the propriety of leaving the press to choose which line they would prefer to have their news by. It would not all do, and as we persisted in dividing our news as requested by the press which we served, we soon after received a notice from the chief operator at the office in New York, Mr. Johnson, that other arrangements had been made, and our news would not be sent, that is, Faxton had engaged to serve a majority of the papers we had been supplying, for less money, and to include the expense of reporting. The line then employed a young man in opposition to us, whom we had instructed and brought up to the business. He was also patronised by F. 0. J. Smith's line in its eastern arrangements, in combination with a portion of the Boston papers, who had defended its course in opposition to the New York Associated Press.

Irregularities also prevailed occasionally on the Morse seaboard line. On one occasion, important California news was started for us at New Orleans, in advance of all others by half a day, which never came to hand; while the same news reached the opposition party in due time. At another period, the burning of the St. Charles Hotel, in New Orleans, was put in for us at a late hour on a Saturday night, soon after which the wires became involved in the flames. Our dispatch was the only one which left the city that night concerning the fire; and communication was not re-established until about 12 o'clock next day; yet, strange to say, our opponent, alluded to above, not only received the news in advance of us, but had it re-forwarded over Smith's line, and published in the Boston papers in advance of us. Not only so, but some members of the press in that city, who patronised us, wrote angry complaints for our having been beat with that important news! Our difficulties were equally as great in getting news from New York to New Orleans. So much so, that the Merchants' Exchange, which we were serving, abandoned the line altogether, and requested us to send their messages by the western or O'Reilly lines; and, if it failed to work, to send none; by which course our business was seriously damaged. We could name other instances of similar management. We appealed to the President of the Morse New Orleans line for an explanation; but no satisfactory solution ever came. We appealed to the Association, and desired that such proceedings should be made public. They seemed duly impressed with the injurious consequences of such management, and caused us to make renewed complaints to the President, promising, if no satisfactory answers were obtained, they would then publish an account of the matter. Nothing further was developed or explained, and the subject was dropped; soon after which we retired, as previously stated, from the general news department. We could on some occasions have justly appealed to the courts of law for redress; but this we declined to do.

From the facts stated in connection with the management of some of the principal Morse lines, it can be readily imagined to what tyranny the press and the people of the United States might have been subjected, had the claimants under the Morse patents succeeded, by expensive lawsuits, in beating off all competition, and securing in their own favour an unrestrained and uncontrolled monopoly!

As a further illustration as to the method in which business

Connection of the Press with the Electric Telegraph

was conducted by some of the lines at that time, we give extracts from an address made by Marshall Lefferts, Esq., President of the Merchants' Line, to the stockholders, in 1851:

As to our Southern and Western business, it has fallen off because of our inability to perform the obligations either with profit or credit to ourselves. We found it necessary to issue a notice to merchants, stating that we would use our best endeavours to forward dispatches beyond the terminus of our own line, but could in no way guarantee their transmission beyond. I must explain, Gentlemen, the nature of this: A. presents himself at our Boston office to send a message (for instance) to New Orleans. We receive and send it to New York, and there hand it over to one of the Southern lines, paying them at the same time the price of transmission for the whole distance, we simply deducting for our service performed. And so the message is passed on, either to stop on the way, or by good luck to reach its destination. If it does not reach its destination—and which is of such frequent occurrence—the sender of the dispatch presents himself at the counter of our office and demands the return of his money. After giving us on the spot the most undoubted evidence of the fact of the message never having reached its destination, we inform him we will make inquiries, and if we can learn which line is at fault we will return him his money. We make inquiry, and when I tell you we can get no satisfaction, it is almost the universal answer; for they all insist on having sent the message through. We explain to the sender of the dispatch, and while he protests, most justly, we are not in a position to refund him his money; for you will bear in mind that we have received but a moiety of the amount for our service, and to refund the whole would be to give back not only what we had fairly earned, but also that which the other lines had received and pocketed, without performing any service. Of the large number of such mistakes and miscarriages which have come under my notice, we have not been able to refund one out of ten. Now what is the effect of this gross mismanagement of the public interest? No argument, no explanation—for none can be made in equity or justice—can satisfy the individual who has been the loser, to a greater or less degree. He has not studied how many lines are to be the bearer of his dispatch—he paid the money to us for its transmission, and although we have performed our obligation as stated by our notice we agreed to do, and he has no legal claim upon us, yet I say he has a moral claim, and failing in its enforcement, he becomes disgusted with both our line and the system. It is true that he still finds himself compelled to use it, but I ask you what is the difference between merchants or individuals using it by compulsion, to keep pace with the times, or his being able and happy to use it as a convenient and reliable means of executing his business? The application I mean as general—and what is the effect produced? On the other hand, what would be the effect, could some of the glaring abuses which clog and fetter the enterprise be removed? I told you we had avoided the Southern and Western business, and that our receipts were less in consequence from this source; but the reason you now know, and I trust appreciate. Better that we should not send a single message, than do it at the expense of our reputation and our conscience.

Some few weeks since I received from Mr. Craig, at Halifax, the indefatigable agent of the Associated Press, a letter touching upon this subject, and I quote from two or three of the letters which passed between us:-

New York, 26th Oct. 1850.

D. H. Craig, Esq., Halifax.

Dear Sir, — Mr. O'Reilly has handed me your letter of the 11th inst. respecting the working of the line between this city and Halifax, and instancing some mistakes, errors, or wilful neglect. It is needless for me to say that *I most heartily agree* to the plan proposed by Mr. Gisbourn, i. e, that the line in fault shall pay the other lines which have done their work correctly, and by which means also the sender of the message will get his money refunded. I have been endeavouring to get all the lines to come into such an arrangement, and still hope for success. I have further, in two or three conversations with Messrs. Hudson and Andrews, urged them to use their influence to bring about such an arrangement, or any rules and regulations to govern the lines, which will put a stop to this plunder of the people. You may build line after line, and cover the United States with this network of wires, but neither the business of Telegraphing

Connection of the Press with the Electric Telegraph

nor its convenience to the public be appreciated, till they can get in return for their money, other than carelessness, errors, and insults; and I know of no surer way of reaching the evil as a beginning than the plan you propose; and I therefore most willingly accept the proposition of Mr. Gisbourn, and we shall soon see which are the lines at fault. I have, like yourself, lately had occasion to inquire into the loss of a message, but I could get no satisfaction, and how absurd it is to have business conducted in such a manner. I do not intend to say that our line makes no mistakes, or is not in some cases at fault; but I wish to find out the errors, and have them corrected by a more efficient staff of operators, and when we are not at fault, have the blame and expense rest upon the right shoulders. I have seen the Superintendent of the Portland line—Mr. C. Hudson—and he is also quite willing to enter into a similar arrangement.

I would suggest, as a means of arriving quickly at the line at fault, that the office from which the message starts should, in case of complaint that it did not go through, require the original to be returned; thus I hope to see all made satisfactory in that respect.

I shall at all times be most happy to receive your suggestions,

And remain, dear sir,
Yours very truly,
Marshall Lefferts,
Pres't N.Y.&N. E. Tel. Co.

5th Dec, 1850.

(Extract.) Craig to Mr. Lefferts.

I was very much gratified a few weeks since, to receive your obliging and perfectly satisfactory letter, and I am happy to inform you that it was regarded with great favour by the Commissioners of their line. I was especially glad to find that I had not, in my voluntary assurance to the Commissioners, overrated your sense of what constitutes justice between Telegraph Companies and the public. Your views were responded to fully and in every particular by the Commissioners, and they, with my permission, enclosed your letter to Mr. Jardine, of the New Brunswick line, and took that occasion to reiterate their firm resolution to insist upon having an entire reform in the whole system of doing business, and thank heaven we can now send dispatches correctly, or else secure the prompt return of our money.

I hope that those to whom Mr. Craig has referred the subject will not (and I know them to be desirous to render their lines worthy of all confidence) let the matter sleep, but by a well-regulated understanding between the companies between Halifax and New York, set an example which must soon be followed by others. What I propose is that, by agreement, messages not sent through within a given time after they are received—or messages which have had errors made in them, or which, for any fault of the line, fail to be of service to the party in consequence of delay—that in all such cases the money shall be refunded; and that the arrangements for carrying out this agreement be of such a character that speedy justice can be done the applicant by the return of his money; and that the line offending shall pay the full amount refunded, so that those lines which may have performed their work promptly shall not lose by the carelessness of the other. Is there anything unreasonable in this as between the lines? and is it not rational towards the public? Let this single correction be carried out, and it will cause the working of the various lines to be more closely looked after, and those clerks who are careless and indifferent, to be replaced by those more competent and faithful; and therefore the merchant will not only have his money refunded when properly demandable, but there will be added a general guarantee to the reliability of the Telegraphic system.

With your sanction, I propose that an appeal through the public prints be made to all the lines, asking them to join in such an engagement, and, as they consent, let the public know which lines are willing to deal upon just principles, and which not. Clear the enterprise of the shackles which now weigh it down, put it in the hands of men satisfied with a fair remuneration, and alive to its national importance, and who shall set bounds to its extension and usefulness! Travel back, as but of yesterday, only six years, and see the few posts and wires—the cumbersome machinery—its slow and inaccurate mode of writing—and now take up the daily journals, and see their pages teeming with telegraphic news from all parts of the Union; and if a few years have accomplished this, what may we not expect from the next few which are to follow!

Connection of the Press with the Electric Telegraph

I have but one other point to which I ask your attention, but it is nevertheless one of much importance—I allude to the non-intercourse policy—which those lines or companies, known as the Morse lines, have set up. The case stands thus: messages which may be sent from Philadelphia, for Boston, coming to our office—we being out of order, or for any other reason—we take them to the Morse line; they will refuse to receive them, although tendered their regular charge for transmission; and the same rule applies to all their lines throughout the country. Can they sustain themselves before the public when the facts are understood? Why, the same rule applies if any of you were a passenger from Philadelphia, for Boston, and when presenting yourself at the New Haven depot, you are told, "We know you came by the Amboy line from Philadelphia, and we are suing that company for an infringement of patent axles, and you cannot go on our line." Such is the refusal of telegraph messages, and I say it is an outrage upon justice, common sense and honesty.

Their refusal rests simply upon a personal quarrel, with which the public have nothing to do, more than they have with all squabbles from one end of the country to the other; and to us it makes no difference beyond the collateral effect which, of the abuses of the system, it brings upon us by disgusting the public with the system and the manner of conducting the business.

See how it is further carried out on our Buffalo line. We expected to make arrangements with the line to and through the Canadas. An individual (a Superintendent of an intermediate line of about twenty miles in length) said he would rather have all his messages mailed at Troy, for New York, than that they should pass over our line. Such a course might suit his feelings, but it fortunately happens that his messages are the property of others (and his generosity to the Morse line is being generous on others' means), and their voice, sooner or later, will be heard from Maine to Georgia, demanding the correction of these abuses. In the hands of men, regardless of the trust over which they are placed as directors, what can we expect of those whose duty it is to obey, and who should have an example of better faith set them? I will make no particular allusion—I am now speaking generally of the system—for my object is not to raise hostility or personal animosity, but simply to lift my voice as one against the injustice which now attaches itself to the business as conducted, and, by pointing out some of the more glaring abuses, lend my feeble assistance to the welfare of the enterprise.

It is often the case on lines, that when they are out of order, to mail their dispatches to their offices, and thence distribute them; and in one case which I looked into personally, it stood thus: A message was sent from Boston, over our line, for a point beyond New York; we handed it over to the line here, paying them something over $3 for it. Shortly after, complaint was made that the message came to hand on the same day as the letter by mail; and on investigation, I found that the telegraph message had been mailed the same day on which we handed it to them, for which service they saw fit to pocket the three dollars. Surely this is telegraphing with a vengeance. At other times, they are thrown amongst the waste paper of the office, or perchance they are given a more respectable death by pitching them into the stove. These things are no fiction—they are facts—and by no means of rare occurrence. But gentlemen, I have no wish to tire you by such details; but that they have a direct and important bearing upon the success of our company cannot be doubted, and by calling your attention to them, I feel confident your suggestions will lead to good results."

The plan of non-intercourse was commenced by F. O. J. Smith, prior to the erection of any competing lines between New York and Boston. Soon after the rupture between the Morse patentees and O'Reilly, he refused to re-send any dispatches to Boston, which reached New York over the O'Reilly lines from St. Louis, Cincinnati, or from any other point at the West. It was immaterial whether they related to illness, deaths, marriages or business; they had to be mailed to Boston, and some days would often elapse before answers could be returned.

Mr. Smith, in his Congressional Report in favour of granting $30,000 to Morse to enable him to build his line, dwelt upon the advantages the telegraph would prove to the public and to the government, and that the American people were desirous the grant should be made. He also alluded to "the grandeur" of the discovery, which, from "a feeling of religious reverence," "the human mind had hardly dared to contemplate." (See *Vail's Magnetic Telegraph Book*, pp. 77, 78.)

Connection of the Press with the Electric Telegraph

We cannot conclude this chapter, without speaking in high praise of the various reporters who were employed by us, or associated with us, at different times, in various parts of the country, and many of whom are still employed by the Associated Press of New York. Among others, we take pleasure in naming C. C. Fulton, Esq., of Baltimore; William Lacy, Esq., of Albany; George W. Brown, Esq., of Buffalo, since appointed consul to Tangier; Mr. Davidson, of the same city; George Bennett and Richard Smith, of Cincinnati; Eugene Fuller, Esq., of New Orleans; Edward Goff Penny, Esq., of Montreal, and Charles Lindsay, Esq., of Toronto, Canada; Joseph Palmer, Esq., and William Stimson, of Boston; and J. B. Skinner, Esq., of Norfolk, Va.; besides many other excellent men, in other less important localities, with others employed as occasional reporters for conventions, public meetings, &c.

15
FAST METHODS OF TELEGRAPH WRITING
FACSIMILE TRANSMISSION OF MANUSCRIPT, PRINTED COPY, AND FIGURES OF ALL KINDS PROPOSED ELECTRO-MAGNETIC LOCKS

Further improvements in Electric Telegraphs are much wanted, in reference to securing greater speed in telegraph writing, with more permanent and constant communication over the wires.

The first object is of difficult accomplishment. The second is attainable by a greater outlay of capital.

We have seen that it requires a slight period of time, and from one to five motions of the finger, to write a single letter; and that the highest rate of speed, so far obtained by the present modes in use, is only about 80 to 100 *letters* per minute, or about 1000 words *per hour*.

When interruptions occur, messages rapidly accumulate on the hands of the operators, frequently including long and valuable news messages for the press. Under such circumstances, if there could be a method of writing employed, a hundred fold more rapid than the present, when the interruption ceased, or the line recommenced working, the whole could be speedily sent forward, and in time to prove valuable to the recipients, and profitable to the line. At present, in cases of this nature, the transmission of delayed messages is so slow, that, before they can be cleared off, they become useless, being superseded by the mail, or are withdrawn by their authors.

Mr. Bain was the first to propose a fast method of communication. He prepared endless slips of paper, about a quarter of an inch in width. These he perforated with holes and slits, to represent the dots and lines of his alphabet. Common writing paper, when dry, is found to be a very good non-conductor of electricity—hence he broke and closed the circuit of a telegraph line, by merely causing this slip of paper to revolve between the conducting points of his instruments. The electric fluid would pass through all the open spaces or holes in the slip, when drawn forward beneath the point of a conducting wire; but the fluid would be arrested, or the current broken, by the intervening solid parts of the paper. Thus, the holes and slits would be rapidly reproduced on the chemical paper in his machine at the opposite end of the wire. As punching a long slip of paper with lengthy messages was a slow process, Mr. Bain invented a very ingenious punching machine, which drew forward the ribbon fillet of paper, and punched it at the same moment, by merely turning a crank with one hand, and touching a key with a finger of the other, to regulate the spacing of the punched holes, to represent the letters of the message.

Mr. Bain's plan was perfectly successful so far as it went; but it was found that, after the quick receipt in cipher of a long dispatch, it would take about as long to copy it into manuscript, as it would to transmit it by the ordinary mode. Another desirable object to attain in telegraphing would be, to send a facsimile copy of a man's handwriting, with his signature. At present, important and confidential business transactions are seldom confided to the telegraph, from the difficulty of knowing whether a name attached to an order to sell and buy bills of exchange, to accept or pay notes, or to advance money, is genuine. Mistakes in figures or modes of expression are also much feared, and not without reason, as experience in some cases has proven.

Mr. Bain was the first to suggest a plan for making facsimile communications. Mr. Bakewell, of England, also made a similar attempt to do the same thing, by contriving machinery which resembled a modification of that used by Mr. Bain in his experiments. A notice of Mr. Bakewell's plan, with specimens of writing performed by it, was published in the *London Illustrated News*, in 1850.

The mode of operating by each was something after this fashion. Imagine a metallic cylinder about eight or ten inches in diameter, and twenty-four inches long, arranged to revolve slowly on its axis by clockwork, and which works in the circuit of the electric fluid at one end of a telegraph wire; while with the other end of the wire a Bain's instrument is connected, and moved with precisely the same velocity. The smooth outer surface of the cylinder may

Telegraph writing, facsimile manuscript, printed copy and figures

have a message written on it, in a non-conducting fluid or material. The writing may be made to encircle the cylinder in manuscript lines, and conclude with the signature of the author. Now, when this cylinder is made to revolve with its written surface in contact with a fine conducting point, it is clear, that every time the point crosses the writing, the circuit will be broken, and while in contact with the cylinder, between the spaces of the letters, the fluid will pass, and produce continuous black marks on the revolving chemically prepared paper in the instrument at the opposite end of the wire. When finished, the whole communication will appear on the chemical paper, in blank letters.

The great objection to the plans of both Bain and Bakewell, arises from the extreme slowness with which the messages are sent, not even equalling in speed any of the methods now in use.

The subject a year or two since engaged our thoughts, and we devised a plan to accomplish the same object, in a manner which will secure extraordinary rapidity of communication; and yet, so arranged as to produce facsimile manuscript letters and signatures, and with figures of all kinds, beyond the possibility of mistake in transmitting it. Not only so, but to send rapidly printed matter. By our method, we proposed to send from 600 to 1000 words per minute.

Besides, our plan, like Bain's and Bakewell's, can be made to send small maps, plans of houses or vessels, &c, as well as profile likenesses of the human face, or the full length outlines of the human figure.

Such a plan, once put in operation, might be found useful for municipal purposes.

Thus, by having the likeness of a rogue, or a close description of his figure and height, a profile representation of his personal appearance could be sent forward in advance of his flight, and copies of it dropped at all the telegraph way-stations through the country.

We entered a caveat some time ago in favour of our proposed improvement; but the want of spare capital and leisure to mature it, has prevented us so far from making any attempt to bring it into public use.

ELECTRO-MAGNETIC LOCKS.

It has occurred to us that electricity might be applied to the formation of electromagnetic locks. They could be made in a simple manner, and add greatly to the security of buildings and vaults. We have not space to give our ideas precisely, as to the best mechanical arrangements to secure the end in view; but in a general way we may say, that the object might possibly be accomplished by some plan similar to the following.

A small battery (portable if preferred), might be used to form a strong electromagnet. This, when placed near the end or brought in contact with a strong door-bolt, resting horizontally or perpendicularly, would exercise sufficient attractive force to withdraw it while the electric fluid should be passing; and when the circuit should be broken, a spring might throw the bolt back in its first position, the door being opened in the meantime. A door might be secured with all the ordinary locks in use, and yet inside them all there might be a secret electromagnetic bolt which no key could find, and which could only be known to and removed by those duly initiated, and having the control of the necessary portable battery and electro-magnet, and which might be removed from the building previous to locking the outside doors. Not only so, but wires could, if necessary, be conducted from the lock to any other apartment in the building, or to a sleeping room; and in case an attempt should be made to force a lock, an alarm could be communicated to the apartments by ringing a bell. In another supposable case, a merchant or banker living up town, if disposed, could have wires conveyed from the locks on his vaults or doors in Wall Street, to his sleeping room, which could be made to convey audible intelligence of any attempt made to force an entrance. We do not offer this idea as ever likely to be adopted in practice, but merely to show the great variety of novel purposes to which electricity may be applied.

As to electromagnetic locks, the mechanical contrivances may be diversified to a great extent, and ultimate experience will only show how the thing can best be effected. The simple enunciation of such a proposed improvement or novelty, is all we have to do with the subject at present, reserving to ourselves the right here-

Telegraph writing, facsimile manuscript, printed copy and figures

after, to secure our claims by a patent if we think proper.

New applications of electricity must continue to occur, so long as the mind of man values the truths of science, or seeks to apply them to the progress and happiness of the race.

16
CALCULATION OF LONGITUDE
MUNICIPAL TELEGRAPHS—BLASTING AND
SUBMARINE EXPLOSIONS
FIRE ALARM WHISTLES
PAGE'S AXIAL ELECTRO-MAGNETIC ENGINE

The first experiments made to ascertain longitude by the use of the electric telegraph, were tried in the United States. As the electric fluid passes in an instant over the longest wire that can be built, even should it "girdle the earth," it is evident that it must beat the flight of time. The revolution of the earth causes fifteen degrees on its surface to pass under the sun's meridian in an hour. When it is twelve o'clock in New York, it is only eleven A.M. at a point fifteen degrees west of it. So that if a dispatch should leave New York at twelve o'clock, it would be received at fifteen degrees west of it, one hour before twelve o'clock. Between New York and San Francisco about fifty degrees of longitude intervene, making a difference of time equal to three hours and twenty minutes. Hence, were a message sent from New York at twelve, and could it reach San Francisco at once, it would beat time three hours and twenty minutes, or arrive there at 8 h. 40 m. A.M. If a similar current could pass round the world without interruption, to the point of beginning, it would of course beat time a whole day or twenty four hours; passing 360 degrees of the earth's surface in less than one second of time.

If a delicately constituted clock with a hand to denote seconds, and its divisions in hundredths, be placed in a telegraph office before an observer in New York, and a similar clock be placed at the other end of a telegraph line in St. Louis, with an observer there, it is only necessary to make the closing and breaking the circuit in New York correspond with the movements of the second hand of the clock in St. Louis (or a good chronometer may be used by each party), to denote the difference in time between the two places in seconds, or fractional seconds, which would indicate with great precision the exact difference of longitude between the two cities. Mr. Lock, of Cincinnati, contrived some ingenious machinery for the purpose of measuring seconds of time, with its divisions, to be recorded by the electric telegraph on fillets of paper, with a view of ascertaining longitude. It was first tried on the telegraph line between Cincinnati and Pittsburgh, and was said to work successfully.

The officers of the U. S. Coast Survey next made experiments on long lines of telegraph wires, for the purpose of determining longitude, and which proved very satisfactory. They made a report of their experiments in 1848. They state that they regard "The value of a night's work with a transit instrument, by the printing method, as about ten times greater than by the method now in use among astronomers." "This year" (1848) "we made abundant experiments on the line from Philadelphia to Louisville, a distance in the air of 900 miles, and in a circuit of 1,800 miles. The performance of this long line was better than any of the shorter lines has hitherto been."

"Not more than two or three good astronomical nights at Cincinnati were lost, by the failure of any part of the line, in the period of two months of our stay at Cincinnati. I learn from an authentic source, that the same success attends the work from Philadelphia to St. Louis—a distance of circuit one-twelfth of the earth's circumference. Great as this distance is, an attempt is to be made to exceed it as soon as circumstances permit, on the line from Halifax to New Orleans, in determinations of longitude."

Lieut. Charles Wilkes, of the late U. S. Exploring Expedition, was the first to experiment with the electric telegraph to determine longitude, which he did in 1844, on the Morse line between Washington and Baltimore, gauging the time by chronometers. The result was, that he ascertained the Battle Monument in Baltimore to be 1' 34.868" east of the Capitol. — (See *Capt. Wilkes's Letter* in Mr. Vail's Book on the Magnetic Telegraph, p. 60.)

The idea of applying an electric telegraph to the determination of longitude, must have been almost coincident with the thought of producing such communication at all; yet it is gratifying to know that the first, most extensive and successful experiments on the subject, have been made in the United States.

Calculation of Longitude, Fire Alarms and Submarine Explosions

The subject we find has, since the experiments made by the members of the U. S. Coast Survey, attracted much attention in England. In Chambers' Papers for the People, vol. ix., 1851, the author says:

> This method of observing is regarded by the Astronomer Royal as of so much importance, that he proposes its use at Greenwich. In discussing this subject before the Astronomical Society, he explained that, "In ordinary transit observations, the observer listens to the beat of a clock, while he views the heavenly bodies passing across the wires of the telescope; and he combines the two senses of hearing and sight (usually by noticing the place of the body at each beat of the clock) in such a manner as to be enabled to compute mentally the fraction of a second when the object passes each wire, and he then writes down the time in an observing book. In these new methods (electric), he has no clock near him, or at least none to which he listens: he observes with his eye the appulse of the object to the wire, and at that instant he touches an index or key with his finger; and this touch makes, by means of a galvanic current, an impression upon some recording apparatus (perhaps at a great distance), by which the fact and the time of the observation are registered."

The experience hitherto obtained of the new method, shows that, in what are termed "irregularities" in observation, the amount "is only about one fourth" of that which occurs with the old method; whether because the sympathy between the eye and the finger is more lively than between the eye and the ear, remains to be determined. The Astronomer Royal proposes to use the "centrifugal or conical pendulum clock," as an instrument superior in every way to those used in America; and "considering,"' as he states, "the problem of smooth and accurate motion as being now much nearer to its solution than it had formerly been, it might be a question whether, supposing a sidereal clock made on these principles to be mounted at the Royal Observatory, it should be used in communicating motion to a solar clock. It might by some persons be thought advantageous, even now, that the drop of the signal ball (at one hour Greenwich mean solar time) should be effected by clock machinery; and it is quite within possibility that a time signal may be sent from the Royal Observatory to different parts of the kingdom, at certain mean solar hours, every day by the galvanic current, regulated by clock machinery."

We formerly suggested in the Journal of Commerce (1848), that balls might be made to drop, or cannon fired in all the principal ports of the United States, by means of the electric current, the moment the sun attained meridian at the National Observatory at Washington, by which chronometers could be regulated at various places, at the same time daily.

At Boston the true time is said to be received every day from the Cambridge Observatory, four miles distant, for the service of the shipping in the harbour.

It has been proposed in England, to apply the telegraph to measure the fluctuations in the barometer and thermometer, by sending up a balloon prepared with instruments, in connection with a galvanic wire, the ground end of which is connected with a recording instrument. It is said an interesting experiment was made in reference to this subject by Mr. Smee, chemist to the Bank of England.

The mysterious influence of the aurora borealis on telegraph lines has been noticed in this country, and also by observers in Europe; among whom was M. Matteucci, who noticed its effects on the apparatus of the electric telegraph line between Ravenna and Pisa, in November, 1850. The following extract from the *Philadelphia Ledger* refers to the effects of an aurora, observed in September 1851. These facts prove the truth of Dr Franklin's original theory, that the aurora borealis is identical with the nature of electricity, or is, in reality, simply an electrical manifestation:

> The aurora borealis, visible on Wednesday and Thursday nights last, was the most brilliant and remarkable exhibition of the kind noticed here for years, and was attended with some very singular phenomena. On Wednesday morning an unusual appearance of atmospheric electricity was manifest on all the telegraph lines radiating from Philadelphia, east, west, and south, which continued more or less till Thursday evening. At times there was a powerful current upon the wires, sustained for minutes, then it would diminish to nothing, and the current from the batteries cease to have any effect on the magnet. It came not in flits and flashes, as is the case during thunder-

storms, but would emit a steady spark for seconds, and even minutes. During this time the weather was cold and remarkably clear. The same effect was noticed in other cities. In Boston, it is said, there was sufficient electricity to supply the telegraph wires without employing the batteries.

MUNICIPAL TELEGRAPHS

Among other important applications of the electric telegraph, is its employment for municipal purposes. It can be used in cities as police station signals, and for giving fire alarms, by ringing bells. The subject has attracted much attention in Boston. Dr W. E. Channing of that city called public attention to it in the columns of the *Daily Advertiser* in 1845. The subject excited no great attention until March 1851, when Dr Channing submitted plans and estimates for erecting municipal telegraphs to the Hon. Josiah Quincy, Jr., mayor of Boston, who recommended their adoption by the city authorities. In June following, the corporation appropriated $10,000 to carry the object into effect.

In 1851 the city of New York connected its eight bell towers with each other, and with the central tower or belfry over the City Hall, by a telegraph wire. This, however, was simply used to signalise an alarm of fire, having no intermediate connections, and had no power to ring fire-bells from the central station.

It is said, that fire alarm telegraphs have recently been erected in Berlin by Mr. Simons, Lieutenant of Engineers, but whether similar to the simple plan adopted in New York, is unknown.

The system for a municipal telegraph, proposed by Dr Channing, is much more comprehensive and complete than any other ever brought to the notice of the public. — (See his account of it in vol. xiii., no. 37, p. 58, of *Silliman's Journal of Science and Arts*, Second Series.)

It is evident that, if various towns, widely scattered over a state or continent, can be brought into speedy communication by the agency of the electric telegraph, that all the central stations of city wards and fire-bells of a large city can be brought into instantaneous connection. With properly contrived instruments, the twenty police stations of the twenty wards and eight bell towers of New York, can be put into speedy and reliable communication at any moment.

Municipal and fire alarm bell electric telegraphs could be, if necessary, extended so as to embrace Brooklyn, Williamsburg, Ward's Island and Harlem, and, if required, be made to include hospital and coroners' calls. They could also be made to embrace all the important points in cities as large as Paris or London. Wherever employed, they will impart great additional power to police forces, as the whole could at any moment be brought to act in concert.

During disturbances in a city like Paris, they would no doubt be found very important aids in the hands of the police, in controlling and suppressing popular outbreaks.

The plan of the Boston Municipal Telegraph described by Dr Channing is quite elaborate, and is illustrated by a number of woodcuts. We have not space to go into details regarding his system, which may for each city be greatly varied. It is sufficient to say, in general terms, that for the contrivance of many of the instruments he employed, he was indebted to Moses G. Farmer, Esq., Electric Telegraph Engineer of that city, and under whose supervision the system was carried into effect. A central office or station is fixed upon, at which the main battery, with other instruments, is placed. From this two circuit wires proceed, like those of the common telegraph wires, fastened to house-tops on ingeniously insulated supports. One of the wires communicates from the main fire-bell tower to all the others, and connects each with machinery, which puts in motion the largest sized hammer, and causes it to strike a large fire-bell the desired number of blows. The other wire proceeds on a still more circuitous route, and from one local street or ward signal station to another. Each station is provided with a strong box, and hinged door and lock. Inside of this box, there is a connecting electromagnet and connecting lever, an axle, with a number of pins in it, to correspond with the number of the station. The axle is turned by a short crank, and in its revolutions the pins break and close the circuit, by moving the end of the lever as often as there are pins or cogs, the result of which is communicated to the central station. If the alarm indicates a fire in the local district

No. 3, the alarm can be instantly rung on all the bells in the city. If it is a subject requiring the speedy and efficient attention of the police, information by alarms can be given at each police station, or the dispatches can be recorded by instruments at each place. The local street alarm boxes are to be placed in charge of a person, whose duty it will be to give the alarm from the local to the central station when called upon, or circumstances require him to do so. The instruments contrived by Mr Farmer for local alarm stations, for isolating wires on house-tops, and for ringing large fire-bells by the action of electromagnetism, in combination with strong clockwork and weights, are very ingenious, and some of which he has patented.

Any number of alarm-stations can be arranged by attaching a corresponding number of pins or cogs, to the axle, in each box, moved by a crank, to break and close the circuit. The time will likely soon arrive when no city will be without its fire-alarm and municipal telegraphs.

Fire Alarm Whistle

We have for some time suggested that a fire alarm, similar to a railroad whistle, would be better adapted for city use than bells. Instead of steam, they would be worked by the force of condensed air, like Daboll's fog-alarm. Whistles of this description can at all times be heard at a much greater distance than bells, and cost a great deal less. To render them more efficient, and to give them such a sound or tone as to distinguish them from ordinary locomotive whistles—which are now heard in the suburbs of nearly all cities —we propose to change the form of the bell-cap. That is, not only to give them greater diameter, but to increase their length, so as to augment the sound. Instead of metal, the bell-cap might be formed of glass, which possesses a high degree of sonorousness. These whistles could be worked by hand, or by the electric telegraph.

We find whistles have also been alluded to and recommended, by Dr Channing, as fire alarms; but no allusion is made as to the possibility of improving or increasing their sound by partially changing their structure, or by the adoption of more sonorous materials.

Whistles can be sounded as readily by the electric telegraph, as ordinary fire bells; and a plan for thus working them has been proposed by Mr Farmer—a diagram of which has been given by Dr Channing, in his paper referred to. The letting on or shutting off the air, however, from the whistle, by the mechanical action of electromagnetism, can be effected by a variety of methods.

A marine whistle moored over a rock, or shoal at sea, or in a harbour, can also be worked at will from shore (after having the air receiver previously charged—say in daylight, or calm weather), by connecting it with the land by a submarine telegraph wire of any required length. Instead of having a man on shore to give the alarms, they could be produced by a train of clockwork on land, to close and break the circuit, and thus keep the whistle sounding. The number of sounds would indicate precisely the character of the spot thus guarded. Vessels crossing the Banks of Newfoundland in dense fogs, in order to prevent collisions, ought also to carry air whistles to be worked by hand.

Dr Channing estimated the municipal wires of Boston at forty-nine miles in length, and the cost of them already put up at $68.72 per mile; and one single set of machines, including one striking apparatus, at $268.

To erect a substantial municipal telegraph for New York, to embrace police stations and fire-alarms, would probably cost from $25,000 to $30,000. If whistles should be used instead of bells, for fire alarms, the expense would be less. The management of the whole, when in operation, would not cost any more, if as much, than the present defective system. Experience has shown that wires, carried over the tops of houses, are less liable to interruption than when carried through cities on poles.

BLASTING AND SUBMARINE EXPLOSIONS, BY ELECTRICITY

Though not immediately connected with telegraphs, yet the important application of electricity to the explosion of powder, deserves a passing notice. We believe that it was Dr Hare, of Philadelphia, who first proposed to apply galvanic electricity to the blasting of rock—which was some twenty-five or thirty years since; and he indicated the proper kind of instruments to be used for the purpose.

Calculation of Longitude, Fire Alarms and Submarine Explosions

When the safety and convenience of blasting by electricity is considered, it is strange that its employment has not become more universal. It is more certain than the common fuse; and the operator can, in firing his charge, choose any distance he pleases, by merely adding to the length of the wire.

Another more important application of the electric current, in producing explosions, may be noticed. Almost coeval with the discovery of the fact that powder could be exploded on land by electricity, it has been known that it also could be made to explode it under water. And it was not long before experiments were tried, both in England and this country. Old vessels were moored off shore, and blown up by kegs of powder fastened to their bottoms.

A more useful application of submarine explosion has been made to the removal of rocks from harbours and the channels of rivers, which obstruct navigation. In those cases where a proper depth of water is found resting above the rock, a large canister of powder, with electrical wires attached to it, is sunk by weights until it rest on the surface of the rock. The superincumbent weight of the water acts as a powerful lever on the force of the powder, so that, when it is exploded by the electric spark, its greatest force is exerted downwards on the face of the rock, cracking, or rending, or tearing it into fragments. By repeating the explosions, the largest and most dangerous rocks that may have been the terror of sailors, and defied the power of man for centuries, are completely removed! The recent successful demolition of Pot-Rock, at Hurlgate, by Professor Maillefort, is still fresh in the public mind. In the course of his operations, he met with a solitary accident, which, however, was of an unfortunate and fatal character; two of his men having lost their lives. The occurrence, however, was not the fault of the electrical fluid. The Professor, with one or two assistants, were indebted to a *Francis' metallic life-boat*—in which they happened to be seated at the time of the explosion—for the safety of their lives.

In producing ignition by the galvanic current, it is necessary, as in the ordinary telegraph line, to use a receiving magnet, which shall complete the circuit of a local battery, near the spot where the rock is to be blasted. This is represented in Figure 9:

The Electric Telegraph in the United States

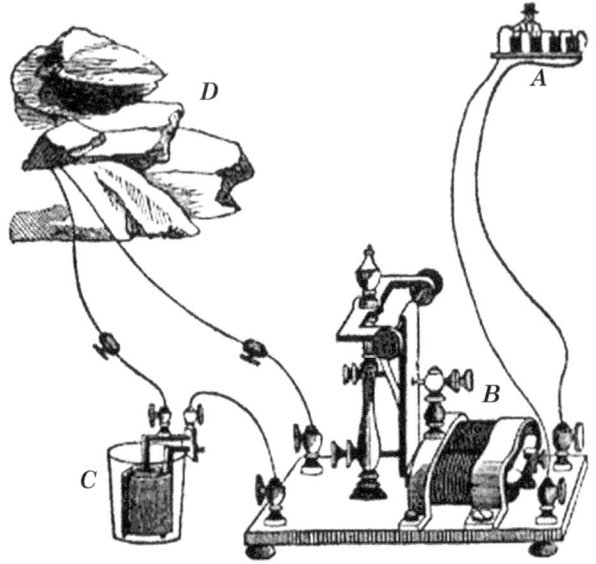

Fig. 9
Rock blasting by the use of an electric current

An operator is seen at a distance ***A*** with the wires extending to a receiving magnet, ***B***, which is constructed on the axial principle, and serves to complete the circuit for the single cup at ***C***, which is to ignite the charge in the neighbouring rocks ***D***.

ELECTRICITY IN WARFARE

In future wars between civilised nations, which we trust are very remote, electricity, like steam, must play an important part. By its agency, not only can intelligence be made to fly from camp to camp, and from army to army, but mines can be sprung—works blown up—heavy ordnance fired—rockets and shells discharged, and submarine powder batteries exploded against hostile fleets. All this it can be made to do, with terrible efficiency and effect.

To blow up vessels in deep water, however, it will require that the charges of powder, unless in immense quantities, should be immersed but a short depth below the surface of the water. A moderate charge exploded on the bottom of a harbour or river, would produce very little, if any, effect on the surface. This we ascertained by experiments made with submarine shells, some years ago, which were exploded by the action of the water, of itself, on coming in contact with it, and without the agency of electricity.

We have also thought it possible that, by a portable small compact electro-galvanic apparatus, combined in some way with the breech or stock of a gun, so as the lock, in cocking and firing, should break and close the circuit, whereby the electric spark given off would discharge the gun, and thus the expense, as well as the inconvenience, of handling the present percussion caps would be got rid of altogether.

No doubt exists that any number of cannons can be instantaneously fired by the electric current. Such an arrangement would enable a 74 gun ship, if desired, to pour her whole broadside, at the same instant of time, into an enemy's ship of war and, moreover, the whole of them could be set off by the finger of a single man, depressing a key to start the electric flash!

We might be expected, when alluding to new applications of electricity and electromagnetism, to refer to the present state, progress and prospects of electro-magnetic mechanical power, and especially to the experiments of Professor Page. The design of our work has been to treat of electricity, chiefly in its relations in some way to electric telegraphs. Where we have digressed, it has been to speak of applications reduced to practice, or to make new suggestions for its employment for other and novel purposes.

AXIAL ELECTRO-TELEGRAPH ENGINE

Dr Page has been liberally patronised by the United States' government, Congress having made liberal appropriations to enable him to bring his Axial Electro-Telegraph Engine into use. Although he has succeeded in producing striking and wonderful results, yet it would seem that his machinery is not sufficiently matured to warrant its employment as a reliable and economical power. Some difficulties probably yet stand in the way of success, which future discoveries and experience may overcome. Those who wish to learn more regarding Dr Page's labours will find them described in various scientific publications of the day.

Were we to indulge in digressions, we might name a vast number of other cases in which electricity is usefully employed, not the least of which is that of the Electrotype, or Electro-Metallurgy; but such digression would be out of place in a work of this description.

We here give a second engraving of House's Printing Telegraph (Figure 10), which exhibits some parts of its ingenious mechanism, not visible in the first.

Fig. 10
House printing telegraph

17
FOREIGN ELECTRIC TELEGRAPHS
THEIR RISE AND PROGRESS IN EUROPE
EXTANT OF LINES, PLANS OF CONSTRUCTION
AND METHODS OF OPERATING THEM

We have found some difficulty in procuring foreign publications which treated fully of the history of Electric Telegraphs, and especially in reference to their present extent and condition in Europe. The fullest account we have been able to obtain, is contained in *Chambers' Papers for the People*, vol. ix., published in the autumn of 1851.

As far as our researches have gone regarding European accounts of the present extent and condition of Electric Telegraphs, they appear very deficient in respect to their knowledge of their rise and progress in the United States.

We were surprised on reading the article referred to, in *Chambers' Papers for the People*, and which purports to be a kind of regular history of the Electric Telegraph, to find the author had omitted all notice of Professor Joseph Henry's valuable discoveries in Electro-Magnetism; and that no mention is made even of his name. Moreover, the valuable and exceedingly ingenious Printing Telegraph of R. E. House is nowhere referred to.

It was not to be expected, perhaps, that a foreign author should be well acquainted with the extent, cost and management of our lines; but, as House had taken out patents in Europe, and had had his machine in practical operation in the United States from two to three years before the publication of the paper referred to, it is strange his invention was not alluded to. Professor Henry's valuable experiments in Electro-Magnetism were published in *Silliman's Journal* as early as 1831, and had been before the scientific world for about twenty years; yet no allusion is made to him or his discoveries, in Chambers' *Electric Communications* or *History of Electric Telegraphs*.

We take the following extracts relating to Electric Telegraphs in Europe from Chambers' work previously referred to. He commences his account of the first Electric Telegraph in England, on the Blackwall Railway, which we saw in operation while on a visit to London in 1840. And in the same year we saw another line, which was built by Cooke and Wheatstone, extending along the Great Western Railway, from London to Slough, near Windsor, about twenty-one miles in length. The first line erected in this country, as we have seen, was in 1844.

    The first application of the electric telegraph was made on the Blackwall Railway, from the station in the Minories to Brunswick Pier. On this line the trains start every quarter of an hour, and the stopping places are so numerous, that it is not easy to conceive how the service could have been performed without such aid as the new mode of telegraphing was calculated to afford. The announcements of departures, of stoppages, of the number of carriages attached to the wire rope, accidents, or other causes of delay, were regularly transmitted, and the business thereby maintained in full vigour and discipline. After this, other railway companies availed themselves of the same indispensable agency, and telegraphs were gradually stretched along the London and North Western, London and South Western, South Eastern, and Eastern Counties lines. On the Great Western the wires at first were placed inside a continuous tube, fixed a few inches above the ground at one side of the way, but were afterwards strained on posts, as on other railways—an arrangement with slight exceptions, now prevalent throughout Britain. This line had not long been complete when a striking instance occurred of the service which the telegraph might render to society. A man of respectable exterior took his seat in a first class carriage at the Slough station, eighteen miles from London: he was a murderer hurrying away from the yet warm body of his victim; the panting engine nears its destination; the eager criminal believes his escape certain; but the alarm has been given at the fatal spot, and quick as lightning the telegraph transmits it to Paddington, with a description of the suspected individual. In three minutes an answer announces the arrival of the train, the identification of the fugitive, and the certainty of his capture. There are few persons who will not remember the impression made on the public mind by this victory of science and justice over crime. Again; a communication transmitted from Paddington immediately that the year 1845 commenced, was received at Slough in 1844, the clock at that place not having struck midnight. Though so short a distance, the difference of longitude was sufficient to mark the inconceivable velocity of the electromagnetic current. Swift-footed Time was henceforward to be beaten in the race.

The wire commonly used for telegraphs is about one-sixth of an inch diameter, covered with a thin coating of zinc, or, as it is called, "galvanised," to prevent oxidation. Besides this, it is found that the deposit from damp and dust and other causes affords a very efficient protection. Four miles of such wire weigh a ton. The posts to which it is attached are fixed at from fifty to sixty yards apart—thirty or thirty-two to the mile. To insure perfect insulation the wires are not permitted to touch the posts, otherwise the current would be diverted downwards through the wood, particularly in wet weather. Insulators of various forms, "rings, collars, and double cones," are made of brown stone ware, which of all substances yet tried throws off the wet most readily. A stone-pitcher, after being plunged into water, is seen to retain scarcely a trace of the immersion beyond a few drops on the surface. Even with this material it is sometimes difficult, during dense fogs or heavy rains, to preserve the integrity of the current.

Besides the supporting-posts, there are others called "winding-posts," four to the mile, to which the wires are connected in alternate half-mile lengths, and stretched by means of a screwing apparatus. It is on these posts that the stone collars are used; a sufficient number being attached to each side, the wire is passed through the eye and drawn tight, while to maintain the communication uninterrupted, a loop of wire is affixed to the main lengths at a short distance on either side of the post, round the front of which it passes in a slight curve. To protect the insulators as much as possible from wet, they are sheltered by a sloping wooden roof. The pointed wires seen rising a few inches above the tops of the posts on some lines is a lightning-conductor with its lower extremity buried in the earth. A precaution not unnecessary, as thunderstorms produce singular effects on the lines of telegraph.

One wire only will suffice for the transmission of correspondence between any two places; the making use of a greater number, six, eight, or ten, as may be seen on some railways, is merely for the sake of economy or convenience. It is found better in practice to keep one or two wires distinct for the main termini or points of correspondence—say from London to Derby—than to make them serve at the same time all the intermediate stations. It is an arrangement which helps to simplify the working duties of the office, and to facilitate them also, for with but one or two wires there would be constantly recurring delays and confusion, since while any two places were intercommunicating all the others would have to wait. One of the wires is sometimes employed exclusively for the alarums—that is, to ring the bell at any station with which it may be desired to "speak." Wherever connection is made with an intermediate office, the main wire is cut, and a shackle inserted, and from either side of this a short wire is stretched to the instrument; thus affording means for the passage of a current up or down the line. The same contrivance would be adopted were there but one wire to connect the two extremes of the line; and it is within the bounds of possibility that some invention or adaptation will show that all the required services may be performed by a single conductor.

The wires, when in their place, are connected with the batteries and telegraphic instruments at the respective stations; and here it becomes necessary to consider the construction and mode of action of a battery. The latter may be familiarly described as a wooden trough, from two to three feet long and about six inches wide, divided crosswise into twenty-four compartments or cells—more or fewer according to circumstances—by partitions of slate. Two plates of metal, copper and zinc, alternately, are placed in each cell, in such an order that all the plates of one kind face towards one end of the trough, and all of the other kind to the other end. A small strip or ribbon of copper unites each pair at the centre of their upper edges, forming, as it were, so many curved handles, by which they can be lifted in and out. As soon, then, as the remaining vacant space in each cell is filled with an acidulated fluid the action commences; the acid begins to act on the zinc by dissolving it, the water contained in the solution is decomposed, and hydrogen thrown off from the surface of the copper plates; while by a combination of oxygen, oxide of zinc is formed, and this, dissolving in the acid—which is commonly sulphuric—sulphate of zinc is produced. These effects are the consequence of the general law established in relation to voltaic electricity, that by the simple contact of dissimilar metallic bodies, a partial transfer of the electric fluid from one to the other invariably takes place. A positive current is generated at the zinc, and passes to the copper through the intervening fluid in all the series of cells; and continues to flow as long as contact is maintained between the wires which depart from either end, whatever be their length. There are various contrivances for increasing and rendering continuous the power of batteries, and for checking deterioration in the metal or acid, which we need not stay to consider, as they do not affect the main question.

The cells of telegraph batteries, instead of a fluid, are filled with pure sand—a material chemically inert, moistened by pouring in the dilute sulphuric acid—an arrangement which admits of the apparatus being removed from place to place without risk of spilling the contents, while it diminishes waste of the plates without diminishing their power. The zinc is most liable to dissolution, and would be rapidly exhausted were it not for the protective influence discovered by Mr. Sturgeon. Having washed the plates clean, he dipped them into mercury, and the thin adherent coat of the rarer metal is found to prevent effervescence of the surface. Those which are known as amalgamated plates consequently last longer than others left in their native state; and after a turn of service they may be again washed and re-dipped. A well-prepared battery, with occasional renewals of the acid, will maintain an effective working condition during twelve or fifteen months.

According to Mr. C. V. Walker, to whose work we are indebted for the substance of some of our details, '"The telegraphs on the South-Eastern Railway, of 180 miles and forty-seven stations, are worked with 2,200 pairs of such plates: and the whole telegraph system in the United Kingdom employs about 20,000 pairs."

In preparing the batteries, it is possible to determine mathematically beforehand the amount of resistance, and the force necessary to overcome it; and thus to proportion the number of plates to the distance to which the wires extend. Large wires are better conductors than small ones. Iron is a better conductor than copper, and copper than silver. The several conditions may be calculated from the formulae laid down by Ohm.

The wires of the battery meet those of the telegraph in what is called the electro-magnetic machine, which externally resembles a cabinet clock, having a square dial-plate inscribed with the letters of the alphabet, and certain arbitrary characters, and two hands placed side by side near its centre. These hands are the needles which are the tongues of the apparatus; in their vibrations to the right and left, their starts and pauses, the whole correspondence is conveyed. For each needle visible on the face of the instrument there is a corresponding one inside, the two being so placed that the north pole of the one and the south pole of the other are in the same position, so as to neutralise their magnetism, or rather the action of the magnetism upon them. They are thus kept in a perpendicular position, and obedient to the slightest impulse from the battery. The inner needle is suspended within a coil or multiplier, which intensifies the current at this particular spot, and is deflected to either side at pleasure by movement of the levers or handles which close or open the electromagnetic circuit.

The telegraph wires finish in two terminals, which form part of the mechanism, and are in connection with the magnet and the multiplier. The battery wires are brought to two other terminals, connected also with the same apparatus; so that in order to reach the telegraph wires, the current must first excite the magnet and the needles. This action takes place only when work is to be done; at other times the circuit is left open. Instantaneously, however, on making contact, the signals exhibited at one end of the line are reproduced at the other; such is the astonishing power of the magnet when rendered active. Messages of business or friendship, congratulation or anxiety, may be sent from one end of the kingdom to the other with the velocity of lightning; on which Arago observes, "the most extended and brilliant flashes of the first and second order, those even which appear to develop their fires over the whole scope of the visible horizon, are not equal in duration to the thousandth part of a second."

When a message is to be sent, the clerk whose duty it is to work the instrument, places the written document before him; and after striking the "ringing key," to call the attention of his correspondent, takes one of the levers which project from the base of the machine in each hand, and moving them from side to side produces corresponding and simultaneous movements of the needles on his own and the distant dial-plate, and the

Foreign Electric Telegraphs—Their Rise and Progress

words are spelt off with great facility. Such is the quickness of apprehension acquired by practice, that the clerks can write the message as fast as the needles deliver it; and it is said that some of the more expert would be able to read it without error from a blank dial.

To expedite transmission, the communications are made as brief as possible, by the elision of letters, and syllables, and sometimes of half a word; besides which, many conventional signs are made use of. "We have," says Mr. Walker, "a signal for the period or full stop and for paragraphs; and we have one for underlining words. And we have many very valuable special signals. There is also a signal among the clerks for laughing, and one for the whistle of astonishment." Where secrecy is desired, any two parties have only to agree to employ numerals as letters, or to reverse or transpose the alphabet at pleasure, in order to form a code of signals which none but themselves shall be able to interpret. The messages transmitted on the Admiralty service are based on a private system, of which the chiefs alone understand the import.

With respect to communications of greater length, the writer just quoted observes: "The rates at which newspaper dispatches are transmitted from Dover to London, is a good illustration of the perfect state to which the needle-telegraph has attained, and of the apt manipulation of the officers in charge. The mail, which leaves Paris about midday, conveys to England dispatches containing the latest news, which are intended to appear in the whole impression of the morning paper. To this end it is necessary that a copy be delivered to the editor in London about three o'clock in the morning. The dispatches are given in charge to us at Dover soon after the arrival of the boat, which of course depends on the wind and the weather. The officer on duty at Dover, having first hastily glanced through the manuscript, to see that all is clear to him and legible, calls "London" and commences the transmission. The nature of these dispatches may be daily seen by reference to the *Times*. The miscellaneous character of the intelligence therein contained, and the continual fresh names of persons and places, make them a fair sample for illustrating the capabilities of the electric telegraph as it now is. The clerk, who is all alone, placing the paper before him in a good light, and seated at the instrument, delivers the dispatch, letter by letter, and word by word, to his correspondent in London; and although the eye is transferred rapidly from the manuscript copy to the telegraph instrument, and both hands are occupied at the latter, he very rarely has cause to pause in his progress, and as rarely also does he commit an error. And, on account of the extremely limited time in which the whole operation must be compressed, he is not able, like the printer, to correct his copy.

At London there are two clerks on duty—one to read the signals as they come, and the other to write. They have previously arranged their books and papers: and as soon as the signal for preparation is given, the writer sits before his manifold book, and the reader gives him distinctly word for word as it arrives; meanwhile a messenger has been dispatched for a cab, which now waits in readiness. When the dispatch is completed, the clerk who has received it reads through the manuscript of the other, in order to see that he has not misunderstood him in any word. The hours and minutes of commencing and ending are noted; and the copy being signed, is sent under official seal to its destination, the manifold facsimile being retained as our office copy, to authenticate verbatim what we have delivered.'

On 11th of December 1849, to the great astonishment of the merchants and bankers of Paris, three gentlemen appeared on "Change" in that city, at half-past one P. M., having with them 150 copies of the *Times*, printed and published in London on the morning of the self-same day; and not only did the *Times* contain the Paris news up to noon of the previous day, but actually the closing prices of the Paris Bourse of the previous evening.

The electric telegraph contributed in no small degree towards the accomplishment of this feat. At eight minutes past one A. M., the dispatch of 321 words, and the Bourse prices, equal to 55 words, were delivered into our charge at Dover, having been conveyed thither from Calais in the ordinary mail-boat. In exactly thirty-two minutes—namely, at forty minutes past one—a correct copy of both these documents was handed in to us by the *Times* office in London. This dispatch occupied us eighteen minutes, being at the rate of $17^{5}/_{6}$ words per minute; the Bourse prices, two minutes. In respect to the latter, the rate is high, because the larger portion is anticipated, the mere fluctuations being all that is new. There was nothing extraordinary to us in this, quickly as it was accomplished; indeed, on the following morning

Foreign Electric Telegraphs—Their Rise and Progress

the writer in London was fairly beaten by the telegraph—the words were read off faster than he could make a clean copy of them."

An idea of the amount of telegraphic correspondence on a railway may be formed from the fact, that on the South-Eastern line, "during the three months ending October 17, 1850, 4831 service messages were entered in the Tunbridge books, and 5235 in those at Ashford." And in six months of the same year the profits arising from the telegraph were £116, being at the rate of 5 per cent, per annum, and an increase of 1½ per cent, over the corresponding six months of 1849.

The proprietors of telegraphs inform us that the communications intrusted to them for delivery comprise the whole catalogue of human wants and wishes, business and pleasure, joy and sorrow, friendship and law. On some occasions they have been asked to send a sum of money, or a small parcel along the wire, by individuals, too, whose surprise showed the sincerity of their belief that the instrument could perform what was desired. Games of chess have been played between parties in distant towns—Southampton and London—the moves being flashed from place to place alternately, as fast as they were made. Then the security which the telegraph lends to railway travelling is not the least of its merits: accident and obstruction can at once be made known, and the remedy provided. "On New-Year's Day 1850, a catastrophe, which it is fearful to contemplate, was averted by the aid of the telegraph. A collision had occurred to an empty train at Gravesend; and the driver having leaped from his engine, the latter started alone at full speed to London. Notice was immediately given by telegraph to London and other stations; and while the line was kept clear, an engine and other arrangements was prepared as a buttress to receive the runaway. The superintendent of the railway also started down the line on an engine; and on passing the runaway he reversed his engine, and had it transferred at the next crossing to the up-line, so as to be in the rear of the fugitive. He then started in chase, and on overtaking the other he ran into it at speed, and the driver of his engine took possession of the fugitive, and all danger was at an end. Twelve stations were passed in safety; it went by Woolwich at fifteen miles an hour, and was within a couple of miles of London before it was arrested. Had its approach been unknown, the mere money-value of the damage it would have caused might have equalled the cost of the whole line of telegraph.'"

The promptitude with which detection has followed fraud by the agency of the telegraph is sometimes rather amusing. Mr. Smee relates an instance: "One Friday night, at ten o'clock, the chief cashier of the bank received a notice from Liverpool, by electric telegraph, to stop certain notes. The next morning the descriptions were placed upon a card and given to the proper officer, to watch that no person exchanged them for gold. Within ten minutes they were presented at the counter by an apparent foreigner, who pretended not to speak a word of English. A clerk in the office who spoke German interrogated him, when he declared that he had received them on the Exchange at Antwerp six weeks before. Upon reference to the books, however, it appeared that the notes had only been issued from the bank about fourteen days, and therefore he was at once detected as the utterer of a falsehood. The terrible Forrester was sent for, who forthwith locked him up, and the notes were detained. A letter was at once written to Liverpool, and the real owner of the notes came up to town on Monday morning. He stated that he was about to sail for America, and that whilst at an hotel he had exhibited the notes. The person in custody advised him to stow the valuables in his portmanteau, as Liverpool was a very dangerous place for a man to walk about with so much money in his pocket. The owner of the property had no sooner left the house than his adviser broke open the portmanteau and stole the property. The thief was taken to the Mansion-House, and could not make any defence. The Sessions were then at the Old Bailey. Though no one who attends that court can doubt that impartial justice and leniency are administered to the prisoners, yet there is no one who does not marvel at the truly railway-speed with which the trials are conducted. By a little after ten the next morning—such was the speed—not only was a true bill found, but the trial by petty-jury was concluded, and the thief sentenced to expiate his offence by ten years' exile from his native country."

The Electric Telegraph Company, incorporated in 1846, whose central establishment is in Lothbury, behind the Bank of England, hold a patent right for a term, in part expired, of fourteen years; their charge for the use of it is £20 per mile. The building is amply furnished with all the requisites for telegraph service: and by means of wires laid in tubes under the

Foreign Electric Telegraphs—Their Rise and Progress

surface of the streets, is connected with all but one or two of the metropolitan railway stations, the post-office, the head police station in Scotland Yard, the Admiralty, the new Houses of Parliament, Buckingham Palace, and the latter, by a further extension, are now placed in communication with the Great Exhibition Building in Hyde Park. Besides these, communications are complete with eighty different places in the provinces, including the chief towns and outposts. Electric telegraphs, according to the parliamentary enactment, "shall be open for the sending and receiving of messages by all persons alike, without favour or preference, subject to a prior right thereof for the service of Her Majesty, and for the purposes of the company." A proviso is also made in favour of the Secretary of State, who may, on extraordinary occasions, take possession of all the telegraph stations, and hold them for a week, with power to continue the occupation should the common weal require it. "There have now,'" so runs the company's official circular, "been established in Edinburgh, Manchester, Liverpool, Glasgow, Hull, and Newcastle, Subscription News Rooms, for the accommodation of the mercantile and professional interests, to which is transmitted by electric telegraph the latest intelligence, including—domestic and foreign news; shipping news; the stock, share, corn, and other markets; parliamentary intelligence; London Gazette; state of the wind and weather from above forty places in England; and the earliest possible notices of all important occurrences." The rate of charges for twenty words is—1d. per mile for the first 50 miles; ½d. for the second 50; and ¼d. for any distance beyond 100 miles. The lowest charge made is half-a-crown. From London to York for twenty words, the cost would be 9s.; to Edinburgh 13s.; to Glasgow 14s.; and to other places in proportion. The number of miles of telegraph in Great Britain at the present time is about 3,000, which leaves about 4,000 miles of railway unprovided for.

During the last session of parliament a second association was incorporated, to be known as the British Electric Telegraph Company, "for the purpose of telegraphic communication upon a more economical scale throughout the country, and for the purchase and use of patents." The company's central office is at the Royal Exchange; they propose to conform to the American tariff of charges for the delivery of messages; to sell licenses; and establish lines to all the chief towns in the kingdom. One of their projects is to connect Dublin with Belfast, and to cross the Channel from the latter town to Scotland: when completed, the capitals of the three kingdoms will be able to intercommunicate at any moment. And the reduction of charge which may be anticipated from the competition will, it is to be presumed, bring the telegraph more than at present within the means of the general public.

The spread of electric telegraphs in France has been extremely slow: for a long time the government refused to abandon their well-developed system of aerial telegraphs; and when with much reluctance they were induced to avail themselves of the infinitely superior agency of electromagnetism, they stipulated that the signals should still be produced by small instruments, counterparts on a diminutive scale of the apparatus contrived by Chappe. There were, however, too many practical difficulties in the way, and ultimately the absurd condition was withdrawn in favour of machinery similar to that used in this country, the government reserving to itself the exclusive use and control of the lines. In 1845 and two following years, the telegraphs extending from Paris to Orleans, to Rouen, to Lille, and Calais, and the Belgian frontier, and to Versailles, were commenced, and brought into operation. The results were such, that in January 1850 a commission was appointed to inquire further into the subject.

They drew up a favourable report, recommending the formation of additional lines, and the plan of stretching the wires on posts in preference to placing them in tubes underground, and that the telegraphs should be open to the use of the public. Among other economical advantages to result from the further extension, was the saving of locomotive power on railways; for, in accordance with the practice on the French lines, whenever a train was twenty minutes late an assistant-engine was dispatched to its relief from one station to another all along the route—an arrangement which not only involved considerable expense, but liability to accident also. The construction of seven telegraphic lines was recommended; five of the number have been officially authorised—from Paris to Tonnerre, Rouen to Havre, Paris to Angers, Orleans to Chateauroux, and from the same city to Nevers; and by a vote of the Assembly, 717,095 francs are set apart to defray the expenses of the necessary works. To afford the fullest facilities to the government, wires are led from the respective stations in Paris to the Hotel of the Minister of the Interior, where the office is now open to the public from 8 A. M. to 9 P. M. every day without exception. Three hundred and one dispatches were transmitted in March, the first month of opening. According to the scale of charges—to send a message of twenty words 62½ miles will cost 3s. 3½d., and 12s. for 620 miles. Two hundred words for the same distances respectively will be 16s. 5d. and 58s. 9d. At this rate to send a message of 300 words from Paris to Calais (195 miles) would cost more than 35s. The commission state, that from seventy-five to eighty letters may be transmitted per minute. In the course of their report they suggest, that as the line from the capital to Dunkerque is on the meridian of Paris, and one of the points of the points of the great survey for the measurement of an arc of the meridian some fifty years ago, the establishment of an electric telegraph will afford an excellent opportunity for testing the former by remeasurement. The telegraphs complete and in progress in France are about 1,500 miles in length.

In Belgium, a commission was also appointed at the close of 1849 to consider the same subject; the individuals named—one of them being M. Quetelet—were eminently qualified for their duties. After a careful examination of the systems of electrotelegraphic communication employed in other countries—the burying of the wires underground, as in Prussia, and the stretching of them on posts, as in England and the United States—the liability to accident from premeditated mischief, atmospheric or other causes—they have decided in favour of wires above rather than below the earth. They show that the disturbances to which the apparatus is liable from electricity of the air is nowhere so effectually guarded against as in England, where conductors are attached to the posts and to the machinery in the offices, and recommend the adoption of similar means of protection on the Belgian lines, which they propose to establish from Brussels to Quievrain—and to the Prussian frontier; from Malines to Ostend by way of Ghent—and to Antwerp—the several distances amounting to about 300 miles. They estimate the annual receipts and savings from these various sources at 86,000 francs; and acting on their report, the government has granted a credit of 250, 000 francs for carrying the project into execution. The central situation of Belgium with regard to other countries renders the formation of these lines of essential importance in continental communications.

Already the ramifications of electro-telegraphs extend from one end of Europe to the other; the lines to connect Petersburg with Moscow, and with the Russian ports on the Black Sea and the Baltic, are in progress; other wires stretch from the capital of the Czar to Vienna and Berlin, taking Cracow, Warsaw, and Posen on the way. Two lines, by different routes—Olmutz and Brunn—unite Vienna with Prague, from whence an offset leads to Dresden; a third enables the Austrian government to send messages to Trieste—their outpost on the Adriatic—325 miles distant; a fourth communicates with the metropolis of Bavaria; and since the 10th January (1850), the *Gazette d'Augsburg* has published the course of exchange in Munich twenty minutes after it has been declared in Vienna. Calais may send news to the city of the Magyar on the Danube, and ere long intelligence will be flashed without interruption from St. Petersburg to the Pyrenees. Tuscany has 100 miles of telegraph under the direction of Signor Matteucci; and a single wire, traversing the level surface of the Netherlands, unites Rotterdam with Amsterdam. Communities are learning that the electric telegraph is an essential of good government; that police without it are inefficient; that by it the interests of humanity are promoted. There is talk also of introducing the thought-flasher into that land of wonders—Egypt; to stretch a wire from Cairo to Suez for the service of the overland mail. Who shall say that before the present generation passes away, Downing Street may not be placed in telegraphic rapport with Calcutta?

In Austria there are about 3,000 miles of telegraph, one-fourth being gutta percha coated wire laid underground. Germany has 3,500 miles complete, and 1,200 more in process of construction. The Austrian government steamboats are fitted with an electric telegraph for communications from the captain on deck to the engine-room.

In a time when mechanical science scarcely admits the signification of "impossible," the insular position of England would not long shut her out from a union with those continental ramifications which we have noticed. The possibility of establishing the connection was satisfactorily proved in August 1850, when a telegraph wire was sunk across the Channel from Dover to Cape Grisnez on the French coast. On the 28th of that month, after certain preliminary experiments had been tried, the *Goliath* steamer started with a huge reel containing 25 miles of wire, coated with gutta percha, on her deck, which was slowly unwound and submerged as she left the land. A horse-box was set up on the beach, to serve as a temporary office for the instruments and operators; from which the wire was led through a lead pipe to some distance beyond low water mark, as a measure of protection in a part the most exposed. A line of buoys marked the track of the steamer; she travelled about four miles an hour, and the wire was gradually sunk at the same rate by means of heavy weights attached at regular intervals. A powerful set of batteries had been provided, as one of the objects was, if possible, to work Brett's printing telegraph; and when the steamer had made good a portion of her voyage, the communication was established, and words were printed at the instrument on board the vessel—imperfectly, it is true; but the fact once verified, the perfecting becomes matter of detail. The needle instrument played freely, and in the evening its signals showed that the voyage had terminated successfully. A message flashed from under the sea by the opposite party pronounced, "We are all safe at Cape Grisnez," with the inquiry added, "How are you?" Thus the international communication was complete; but soon after interrupted by the breaking of the wire, which was too weak to withstand the action of the water and friction on a rocky bottom.

As before observed, the possibility having been proved, the Submarine Telegraph Company, whose patent embraces England, France, and Belgium, set about preparations to re-establish the connection on a scale calculated to obviate the risk of accident. The wires, four or five in number, are to be enclosed in cables several inches in thickness, and from twenty to twenty-five miles in length, each weighing 490 tons. It is proposed to have three or four such cables, to be anchored to the bottom two or more miles apart, so that if one should fail, communication may still be maintained by the others. Expectations are held out that the line may again be brought into working order during the present year (1851.).

Since then, by an improved telegraph wire or jointed iron cable stretched across on the bottom of the Channel, communication has been regularly and permanently established, by which means London and Paris are in daily, and almost instant communication. The aggregate length of telegraph wires now in operation in America and Europe, exceeds the circumference of the earth's surface; and in a few years will probably more than double this distance.

18
PROGRESS AND OPERATION OF FOREIGN TELEGRAPHS
FINANCIAL RETURNS OF ENGLISH TELEGRAPHS MERIDIAN TIME BY ELECTRICITY—SUBMARINE TELEGRAPHS—CONTINENTAL LINES–CONCLUSION

Since the publication of the article on Telegraphs in *Chambers' Papers for the People*, this past year (1851), we have gleaned some additional information from late English papers.

The particulars of these addenda, though brief and meagre, nevertheless plainly indicate the rapid spread of Electric Telegraphs over the civilised world.

FINANCIAL RETURNS OF ENGLISH TELEGRAPH LINES

"The half yearly meeting of the proprietors was held on Saturday, the 29th Nov. 1851, at the central station in Lothbury, the Chairman, Mr. J. L. Ricardo, M. P., presiding.

"The accounts show an available balance of £14, 701 12s. 3d. Of this sum the directors recommend the division of £9,369 9s., which will produce a dividend at the rate of six per cent, per annum upon both classes of shares, leaving a balance of £5,332 3s. 3d. to the credit of the next half year. The capital account, to the end of December last, showed a receipt upon the shares of £330,000; sundry liabilities, as per ledger, £34,981 19s. 9d.; reserved fund, £68,534 16s. 9d.; balance, £14,701 12s. 3d.; total £428,218 8s. 9d. On the other side, the cost of telegraphs, completed and in progress, inclusive of the cost of patents, was £361, 731 18s. 8d., other items (including £26,370 19s. 7d., cash and securities in hand), £76,486 10s. 1d.; total £448, 218 8s. 9d. The revenue account for the half year ending December 31 showed receipts for messages, subscriptions and contracts, amounting to £24,336 8s. 10d.; and expenses amounting to £15,979 16s. 7d.; leaving a balance of £8,681 12s. 3d. The report having been received and adopted, formal resolutions carrying out its several recommendations, and making a dividend at the rate of six per cent., were also agreed to; Mr.

George Wilson and Colonel Wylde were re-elected directors, and Mr. Albert Ricardo was re-appointed an auditor."

MARKING MERIDIAN TIME BY ELECTRICITY

In a previous part of our work, we have alluded to the practicability of causing meridian time, at any central observatory, to be simultaneously noted at various other localities, which would not only indicate the true difference of longitude, but regulate the time for each place. By the following account from a London paper (April, 1852), we see that the Astronomer Royal is about to put a similar plan into practical operation. The National Observatory at Washington should, in some similar way, and for a like purpose, be put in electrical communication with New York and other seaport towns.

ELECTRICAL TIME. — The Electric Telegraph Company are now introducing a novel and beautiful system for distributing and establishing correct Greenwich or uniform time throughout the country. For this purpose, telegraphic wires are being laid down over the railway and through Greenwich Park to the Observatory, and, through the liberality of the South-Eastern Railway, the wires are being carried from thence, at the instance of the Astronomer Royal, to the telegraph office in the Strand, on the dome of which, facing St. Martin's Church and Charing Cross, is to be placed an elevated pole, similar to that on the top of the Observatory at Greenwich; from which, every day at noon, a large black ball, four feet six inches in diameter, will, by electro-motive power, be dropped, descending simultaneously to a second with that at Greenwich, both being, in fact, liberated by the same hand; and, after falling on a cushion, a contrivance at the base of the pole, communicating standard time by the wires diverging from Lothbury and the Strand, by an electrical *coup,* throughout the country. This ingenious apparatus is designed by Mr. Edwin Clark, Chief Engineer of the Electric Telegraph Company.

REGULATION OF TIME BY ELECTRIC TELEGRAPH.

In addition to the above, the *London News*, of the 17th April, contains the following interesting details respecting the execution of the enterprise referred to:-

Foreign Telegraphs—Financial Returns, Etc.— Conclusion

The ingenious apparatus intended, by means of the electric telegraph, to establish and transmit uniformity of time, in accordance with the Greenwich standard, throughout London and the provinces, is now completed. The apparatus in question has been designed by Mr. Edwin Clark, the engineer to the Electric Telegraph Company, and carried out under his superintendence by Mr. John Sandys, electric telegraph instrument and clock manufacturer, at whose establishment, in Upper Whitecross street, the contrivance was exhibited to visitors. Entering an extensive workshop, the visitor observes between fifty and sixty workmen busily engaged in all sorts of telegraphic construction, brass finishing, battery plates, wire work, &c.

There are to be seen in this department, in process of manufacture, all the newest devices and designs for carrying out with economy, and to the greatest possible perfection, the system of communication introduced and now established so extensively by the Electric Telegraph Company. In another department were observable a variety of finished instruments ready for removal, and in a third direction the visitor entered a department devoted to the ingenious apparatus constructed by the Electric Telegraph Company, in association with the Astronomer Royal, for the establishment and transmission throughout London and the various provincial and commercial communities of mean Greenwich or geometrical time. In this department, placed horizontally on supports, is a long quadrangular shaft or pillar of wood, painted white without, thirty-eight feet high, and between eight and ten inches square. It is constructed in three sections, so as to allow of its being readily taken to pieces, in about thirteen feet lengths. This hollow shaft passes, towards its apex, through the centre of a large hollow globe or ball, by a contrivance similar to that in connection with the well-known ball at the top of Greenwich Observatory. The globe is about four feet six inches in diameter, is constructed entirely of basket work, and is large enough for four persons to sit comfortably in it. It is to be covered with canvas and painted black. By means of a bar inside the ball, connected with a piston or movable rod, working in grooves or slides from inside the pillar, and by means of an electrical trigger communicating with it, an electrical shock will every day at one o'clock cause the ball to fall eight feet from the top of the pillar, and by this means, when placed in its intended position in the Strand, it will, when the wires connect it with the ball on the top of the Observatory at Greenwich, fall simultaneously with it, and thus indicate exact Greenwich time. The ball, released by the electrical trigger, will on the instant of its falling communicate mean time to a "grand regulator," or electrical clock, in the offices below, which, in its turn, will send an electrical stream of true time throughout the country—the regulator being also furnished with a remarkable apparatus, by means of which the minute and hour hands are to be made to move every minute. The apparatus consisting of the pillar and ball will be placed on the top of a dome admirably adapted for the purpose, and which is observable on the top of the West Strand establishment of the Electric Telegraph Company, by Hungerford market. Of the thirty-eight feet of which the pillar consists, seventeen feet are to be fixed into the dome or roof, leaving twenty-one feet visible above the building, and the pillar will be crested with a silvery-looking weather vane ten feet six inches high, with the letters E T C cut transparently on the arms. Outside the establishment in the West Strand, but in immediate communication with the grand regulator, is to be a newly designed electrical clock, surmounting an ornamented bronze pillar, and similar ones are to be placed at all the other telegraph stations. The dial-plate of these clocks is of the most elegant workmanship and design. There are four square dial faces, each two feet high, and within which the electrical clockwork is enclosed. The dial plates are of a beautiful enamelled glass of a pure milk white, the hours and minutes being marked by stained glass indications on the alabaster base, and upon which, under the hour and minute hands of bronze, are inscribed, in beautiful light blue lettering, the words ELECTRIC TELEGRAPH COMPANY. The peculiarity of the enamelled glass, which consists of a coating of white enamel on both sides of the ordinary sheet glass, is, that when the dial is illuminated it does not reflect on its frontispiece or face the unsightly shadows of the machinery within, so noticeable in connection with the ordinary ground glass illuminated clocks in London, but presents a softly opaque white surface or transparency, and is, so far, a great improvement upon the present method of illuminated clock-making.

Mr. Sandys is the patentee."

Foreign Telegraphs—Financial Returns, Etc.— Conclusion

A new Telegraph line was opened between London and Bristol on the 5th March, 1852.

SUBMARINE TELEGRAPHS

Operations of the Line connecting France and England.

According to a late report made by the Submarine Telegraph Company between England and France, the following statement of receipts and expenses are given. It will be perceived that their receipts nearly doubled in the second month after the line was opened. Should a corresponding ratio of increase for the remaining months of the year be realised, its income over expenses will be enormous.

According to a report of the Submarine Telegraph Company, (between France and England) submitted at a meeting held in Paris, it appears that the receipts were £398 in the first month from the commencement of operations, £517 in the second, and £519 in the third. The annual expenses of all kinds, it was estimated, would not exceed £2000.

Since the successful operation of the submarine telegraph connecting England and France, across the English Channel, a number of other submarine lines have been projected, some of which are noticed in the following paragraphs from foreign papers:

From the London News of April 10th, 1852.

THE IRISH SUBMARINE TELEGRAPH. — The difficulties that have been interposed to the construction of a submarine telegraph between Great Britain and Ireland are at length in a fair way to be overcome. The communication between London and Dublin is expected to be formed by the 20th day of May. Port Patrick and Donaghadee are the points from which it is proposed to throw the wires across the Channel, as the line will then be shorter by forty-four and a half miles than that contemplated between Kingstown and Holyhead. The company propose to lay down two distinct lines of four wires, and they will be in full co-operation with the Electric Telegraph Company established in London. A novel feature of the proposed plan will consist in the connection of the Government offices in Downing Street with the Irish metropolis, an advantage of no little importance. A great benefit will be bestowed on the commercial interests of the country by the facility of communicating with the frequented port of Queenstown, as harassing and expensive delays are now constantly experienced from the necessity of vessels awaiting the receipt of orders for several days in the offing."

A Belgian paper remarks that:

> Some English gentlemen are at this moment engaged in studying plans at Nieuwdiep, as it would appear, relative to a project for the establishment of a submarine electrical communication between Holland and England. If we may give credit to the reports in circulation, this plan would only be carried into execution, in the case of France ceasing to be on friendly terms with England, which circumstance would, consequently, render it necessary, in order to insure the electric communications between England and the continent, to choose other points, situated so as to be protected from the effects of hostilities.

A late English paper gives the following:

> LONDON AND BELGIUM. — The King of Belgium has granted to Sir James Carmichael, Bart., the exclusive privilege of a submarine line between England and Belgium. It is to communicate ultimately with India.

ELECTRIC TELEGRAPH LINES ON THE CONTINENT.

From the 1st of March 1851, to the 1st of February, the French government electric telegraph, of which the head office is established at the Hotel of the Minister of the Interior, has transmitted 11,443 messages, as follows:-

1. Dispatches transmitted from Paris to the departments, from the departments to Paris, and from one department to another, 4,594.
2. Dispatches transmitted from France to Belgium, and from Belgium to France, 4,774.
3. Dispatches transmitted from France to England, and from England to France, 1,468.
4. Dispatches transmitted *in transitu* from England to Belgium through France, and from Belgium to England, 607, making altogether 11,443 dispatches. The sums paid for the transmission of these dispatches amounted to 166,577f.,

Foreign Telegraphs—Financial Returns, Etc.— Conclusion

viz.:- For France, 97,889f.; for Belgium and Germany, 50,322f.; for England, 18,346f.

It appears from the following statement, that nearly one-third of the dispatches transmitted over the Prussian Electric Telegraph lines in 1851 were on government account.

Prussia. — The total length of electric telegraph lines in Prussia is 446 German miles (981 leagues), and 376 of them are underground. 130 of them were established in the course of the last year. The number of. dispatches transmitted in 1851 was 39,972, of which 11,447 were by the Prussian or other governments—the rest by private individuals. The total receipts were 90,450 thalers (343,710fr.), and the total expense 157,162 thalers (597,215fr.); but the government dispatches are not included in the receipts.

Italy. — The Minister of the Interior at Modena has issued a notice, announcing that the telegraphic line which connects Modena, Guarilla and Reggio with the Austrian States, will be henceforward open to the use of the public. The tariff is regulated according to the distance in German miles, of fifteen to a degree, and no dispatch is to contain more than one hundred words.

It is supposed by many that telegraph wires buried in the ground, with no other coating than Gutta Percha, will not prove to be permanent. Some assert that it will rot, like wood, and is liable to be cut or destroyed by burrowing animals or vermin. It is true that wires coated with gutta percha could be encased in iron tubes; but this would greatly enhance the cost. We have thought, that if the wires, thus coated, were afterwards surrounded with a thick mass of pine rosin melted with sand, that the security of the wires would be attained, and in a cheap and durable form. The melted materials could be poured or cast around the coated wire, after it should be properly placed at the bottom of a trench dug in the ground. The rosin and sand forms a mass nearly as hard as stone, and it is at the same time a most excellent non-conductor of electricity, and impervious to water.

Conclusion

It will be seen that we have traced Electricity from an early period, chiefly in its bearings upon Electric Telegraphs. We have followed up the progress of discoveries, which finally ended in the establishment of telegraphic communication, as we now see it. The early triumphs of the Electric Telegraphs are events of the past—our hopes and aspirations look to the future, when their improvement, progress and results will prove more astonishing and truly brilliant than any we have ever yet witnessed. Like kindred discoveries, the science of Electric Telegraphy is struggling through its childhood. Our knowledge of the wonderful agent by which it is actuated is yet imperfect. We only know it by its effects. Future investigations must disclose new discoveries in electricity, applicable to new and important purposes, conducive alike to the advancement of man's power over matter, of his elevation of intellect and higher civilisation.

CANST THOU SEND LIGHTNINGS, THAT THEY MAY GO, AND SAY UNTO THEE, HERE WE ARE? — JOB.

19
CHARGES FOR TELEGRAPHIC DESPATCHES
FROM NEW YORK TO ALL PARTS OF THE UNITED STATES AND CANADA

LIST OF OFFICES AND TARIFF.	Ten words.	Each add'al word.	LIST OF OFFICES AND TARIFF.	Ten words.	Each add'al word.
Adrian, Mich.	$ 95	5	Bedford, Pa.	$ 50	8
Akron, Ohio	65	4	Belfast, Me.	65	4
Albany, N. Y.	30	2	Bellefontaine, Ohio	80	6
Albion, "	40	3	Belleville, Can.	75	6
Albion, Mich.	90	5	Belvidere, Ill.	1 30	8
Alleghany City, Pa.	60	4	Bellville, Ohio	75	4
Alexandria, Va.	56	5	Beloit, Wis.	1 50	9
Algonac, Mich.	1 10	6	Bennington, Vt.	55	4
Allentown, Pa.	40	4	Berthier, Can.	1 10	9
Alton, Ill.	1 50	9	Bethlehem, Pa.	40	4
Amherst, Me.	1 53	9	Binghampton, N. Y.	40	3
Ann Arbor, Mich.	90	5	Bloomington, Iowa	1 50	9
Appleton, Wis.	1 80	10	Bolivar, Ohio	75	4
Ashland, Ohio	75	5	Boston, Mass.	20	2
Ashtabula, "	50	3	Bowmansville, Can.	75	6
Athens, "	1 00	7	Brantford, "	83	6
Athens, Ga.	95	9	Brattleboro, Vt.	45	4
Atalanta, "	1 91	12	Bridgeport, Conn.	20	2
Attica, Ind.	1 10	6	Bristol, Mich.	1 05	6
Auburn, N. Y.	40	3	Brockport, N. Y.	40	3
Augusta, Me.	65	4	Brockville, Can.	90	7
Augusta, Ga.	1 33	9	Brownsville, Pa.	50	4
Aurora, Ill.	1 30	7	Buffalo, N. Y.	40	4
Avon Springs, N. Y.	45	3	Burlington, Vt.	60	4
			Burlington, Iowa	1 35	10
Baltimore, Md.	50	4			
Bangor, Me.	80	4	Cadiz, Ohio	80	4
Baraboo, Wis.	2 00	11	Cahawba, Ala.	1 90	12
Bardstown, Ky.	1 30	9	Cairo, Ill.	1 50	9
Batavia, N. Y.	40	3	Calais, Me.	90	5
Batavia, Ill.	1 25	7	Cambridge City, Ind.	1 10	6
Bath, Me.	70	4	Camden, S. C.	1 03	7
Baton Rouge, La.	2 35	16	Canal Dover, Ohio	1 00	7
Battle Creek, Mich.	90	5	Canandaigua, N. Y.	40	3
Beardstown, Ill.	1 40	8	Canton, Miss.	1 75	11
Beaver, Pa.	70	4	Cantwell's Bridge, Del.	45	4
Beaver Meadow, Pa.	50	4	Carbondale, Pa.	40	3

The Electric Telegraph in the United States

LIST OF OFFICES AND TARIFF.	Ten words	Each addi'al word	LIST OF OFFICES AND TARIFF.	Ten words	Each addi'al word
Carlisle, Pa.	$ 40	3	Delphi, Ind.	$1 00	6
Carlisle, Ind.	1 10	6	Deposit, Pa.	40	2
Carmel, N. Y.	20	1	Detroit, Mich.	75	4
Castleton, Vt.	55	4	Dixon, Ill.	1 35	8
Catskill, N. Y.	20	1	Dodgeville, Wis.	1 60	10
Catasauqua, Pa.	40	4	Dover, Del.	50	5
Cedarsburgh, Wis.	1 50	8	Dover, Ohio	75	4
Chagrin Falls, Ohio.	75	4	Doylestown, Pa.	45	4
Chambersburgh, Pa.	40	3	Dresden, Ohio	75	4
Charleston, S. C.	1 19	8	Dundas, Can.	83	6
Cheraw, "	97	7	Dundee, Ill.	1 35	7
Cherryfield, Me.	90	5	Dunkirk, N. Y.	50	3
Chicago, Ill.	1 00	6			
Chillicothe, Ohio	1 00	7	Easton, Pa.	40	4
Chippewa, Can.	63	6	Eastport, Me.	90	5
Cincinnati, Ohio	75	5	East Machias, Me.	90	5
Circleville, "	1 00	6	Eaton, Ohio	95	6
Clarksville, Tenn.	1 25	9	Eddyville, Ky.	1 35	8
Clarkston, Mich.	1 00	6	Elgin, Ill.	1 30	7
Claremont, N. H.	45	4	Elkhart, Mich.	1 05	6
Cleveland, Ohio	50	3	Ellsworth, Me.	70	4
Clinton, La.	2 00	14	Elmira, N. Y.	40	2
Coburgh, Can.	50	4	Elyria, Ohio	60	4
Cold Spring, N. Y.	20	1	Enfield, N. H.	50	4
Coldwater, Mich.	1 00	5	Erie, Pa.	50	3
Columbia, Pa.	55	5	Eugene, Ind.	1 10	6
Columbia, S. C.	1 06	7	Evansville, Ind.	1 60	11
Columbia, Tenn.	1 40	9			
Columbus, Ohio	80	5	Fairport, Ohio	55	4
Columbus, Ga.	1 75	11	Fall River, Mass.	20	2
Columbus, Miss.	1 75	12	Fayetteville, N. C.	90	7
Columbus, Tenn.	1 75	11	Flint, Mich.	1 05	6
Concord, N. H.	45	4	Fon Du Lac, Wis.	1 70	10
Coneaut, Ohio	50	3	Frankfort, Ky.	1 40	9
Connellsville, Ind.	1 05	7	Franklin, N. H.	45	4
Constantine, Mich.	1 05	6	Franklin Mills, Ohio	75	4
Cooperstown, N. Y.	40	3	Fredericksburgh, Va.	61	5
Corning, "	45	3	Frederick City, Md.	75	7
Cornwall, Can.	85	7	Fredericktown, N. B.	1 28	7
Covington, Ind.	1 10	6	Fredonia, N. Y.	50	3
Coxsackie, N. Y.	20	1	Freemont, Ohio	60	4
Crown Point, Mich.	1 10	6	Freeport, Ill.	1 40	9
Cumberland, Md.	80	7	Fort Atkinson, Wis.	1 50	9
Cuyahoga Falls, Ohio	65	4	Fort Plain, N. Y.	30	2
Cuylersville, N. Y.	45	3	Fort Wayne, Ind.	1 00	6
			Fort Winnebago, Wis.	2 00	11
Damariscotta, Me.	60	4	Fulton, N. Y.	40	3
Dansville, N. Y.	45	3	Fulton, Ohio	75	4
Dansville, Ill.	1 00	6			
Darling, Can.	75	6	Galatin, Miss.	1 90	13
Dayton, Ohio	1 00	6	Galena, Ill.	1 40	8
Debuque, Ill.	1 40	6	Galipolis, Ohio	1 15	8
Defiance, Ohio	95	6	Gardiner, Me.	65	4
Delaware, Pa.	25	2	Geneva, N. Y.	40	3
Delaware City, Del.	40	4	Geneva, Ill.	1 35	7

Charges for Telegraphic Dispatches from New York

LIST OF OFFICES AND TARIFF.	Ten words	Each addit'al word.	LIST OF OFFICES AND TARIFF.	Ten words	Each addit'al word.
Geneseo, N. Y.	$ 45	3	Junction, Vt.	$ 50	2
Georgetown, D. C.	55	5			
Germantown, Pa.	45	4	Kalamazoo, Mich.	90	5
Girard, "	50	3	Kenosha, Wis.	1 30	7
Glasgow, Ky.	1 40	9	Kenton, Ohio	80	6
Goodrich, Mich.	1 05	6	Keokuk, Iowa	1 50	9
Goshen, N. Y.	25	1	Kingston, N. Y.	20	1
Goshen, Ind.	1 05	6	Kingston, Can.	75	6
Granville, Ohio	80	5			
Green Bay, Wis.	2 00	11	Lafayette, Ind.	1 00	6
Greenbush, "	1 60	10	Lake Mills, Ohio,	1 50	9
Greencastle, Ind	1 10	6	Lancaster, Pa.	45	4
Greenfield, Mass.	45	4	Lancaster, Ill.	1 55	10
Greensburg, Pa.	60	4	Lancaster, Ohio	95	6
Griffin, Ga.	1 86	12	Lancaster, Wis.	1 65	10
			Laporte, Ind.	1 00	6
Halifax, N. S.	1 65	10	Lasalle, Ill.	1 40	8
Hallowell, Me.	65	4	Lawrenceburg, Ind.	95	6
Hamilton, Ohio	1 10	6	Lebanon, N. H.	50	4
Hamilton, Can.	63	5	Lewiston, N. Y.	50	4
Hannibal, Mo.	1 50	9	Lexington, Ky.	1 20	8
Hards, N. B.	1 65	10	Liberty, Miss.	1 95	12
Harper's Ferry, Va.	75	7	Lima, Ind.	1 00	6
Harrisburg, Pa.	45	4	Little Falls, N. Y.	30	2
Hartford, Conn.	20	2	Littlefort, Ill.	1 30	8
Havre de Grace, Md.	45	3	Lockport, N. Y.	40	3
Hazlegreene, Wis.	1 60	9	Lockport, Ill.	1 30	8
Hebron, Ohio	80	5	Logansport, Ind.	1 00	6
Herkimer, N. Y.	30	2	London, Can.	88	7
Hillsboro, Ill.	1 30	8	Louisville, Ky.	1 20	8
Hollydaysburg, Pa.	60	4	Lowell, Mass.	45	4
Holly Springs, Ala.	1 90	13	Lowell, Ohio	80	4
Honesdale, Pa.	25	2	Lower Sandusky, Ohio	70	4
Hudson, N. Y.	20	2			
Hudson, Ohio	65	4	Macon, Ga.	1 66	10
Hudson, Mich.	95	5	Madison, Ind.	1 20	7
Huntington, Ind.	95	6	Madison, Wis.	1 50	9
			Manchester, Vt.	55	4
Indianapolis, Ind.	1 10	6	Manchester, N. H.	55	4
Iowa City, Iowa	1 75	11	Mansfield, Ohio	75	4
Ithica, N. Y.	65	4	Marengo, Ill.	1 30	8
			Marietta, Ohio	1 10	7
Jacinto, Miss.	1 55	11	Marshall, Mich.	90	5
Jackson, "	2 20	11	Martinsburgh, Va.	75	6
Jackson, Mich.	90	5	Massillon, Ohio	65	4
Jacksonville, Ill.	1 45	9	Maumee City "	70	4
Janesville, Wis.	1 40	9	Maysville, Ky.	1 00	6
Jefferson, N. Y.	65	4	Meadville, Pa.	70	5
Jefferson, Miss.	1 50	8	Medina, N. Y.	40	3
Jefferson, Wis.	1 50	9	Medina, Ohio	75	4
Jeffersonville, Ind.	1 10	8	Memphis, Tenn.	1 75	13
Jersey City, N. J.	20	2	Menassa, Wis.	1 90	10
Jonesville, Mich.	95	5	Meriden, Conn.	20	2
Juliet, Ill.	1 30	8	Michigan City, Ind.	1 00	5
Junction, Ohio	90	6	Middlebury, Vt.	60	4

199

The Electric Telegraph in the United States

LIST OF OFFICES AND TARIFF.	Ten words.	Each addi'al word.	LIST OF OFFICES AND TARIFF.	Ten words.	Each addi'al word.
Geneseo, N. Y.	$ 45	3	Junction, Vt.	$ 50	2
Georgetown, D. C.	55	5			
Germantown, Pa.	45	4	KALAMAZOO, Mich.	90	5
Girard, "	50	3	Kenosha, Wis.	1 30	7
Glasgow, Ky.	1 40	9	Kenton, Ohio	80	6
Goodrich, Mich.	1 05	6	Keokuk, Iowa	1 50	9
Goshen, N. Y.	25	1	Kingston, N. Y.	20	1
Goshen, Ind.	1 05	6	Kingston, Can.	75	6
Granville, Ohio	80	5			
Green Bay, Wis.	2 00	11	LAFAYETTE, Ind.	1 09	6
Greenbush, "	1 60	10	Lake Mills, Ohio,	1 50	9
Greencastle, Ind.	1 10	6	Lancaster, Pa.	45	4
Greenfield, Mass.	45	4	Lancaster, Ill.	1 55	10
Greensburg, Pa.	60	4	Lancaster, Ohio	95	6
Griffin, Ga.	1 86	12	Lancaster, Wis.	1 65	10
			Laporte, Ind.	1 00	6
HALIFAX, N. S.	1 65	10	Lasalle, Ill.	1 40	8
Hallowell, Me.	65	4	Lawrenceburg, Ind.	95	6
Hamilton, Ohio	1 10	6	Lebanon, N. H.	50	4
Hamilton, Can.	63	5	Lewiston, N. Y.	50	4
Hannibal, Mo.	1 50	9	Lexington, Ky.	1 20	8
Hards, N. B.	1 65	10	Liberty, Miss.	1 95	12
Harper's Ferry, Va.	75	7	Lima, Ind.	1 00	6
Harrisburg, Pa.	45	4	Little Falls, N. Y.	30	2
Hartford, Conn.	20	2	Littlefort, Ill.	1 30	8
Havre de Grace, Md.	45	3	Lockport, N. Y.	40	3
Hazlegreene, Wis.	1 60	9	Lockport, Ill.	1 30	8
Hebron, Ohio	80	5	Logansport, Ind.	1 00	6
Herkimer, N. Y.	30	2	London, Can.	88	7
Hillsboro, Ill.	1 30	8	Louisville, Ky.	1 20	8
Hollydaysburg, Pa.	60	4	Lowell, Mass.	45	4
Holly Springs, Ala.	1 90	13	Lowell, Ohio	80	4
Honesdale, Pa.	25	2	Lower Sandusky, Ohio	70	4
Hudson, N. Y.	20	2			
Hudson, Ohio	65	4	MACON, Ga.	1 66	10
Hudson, Mich.	95	5	Madison, Ind.	1 20	7
Huntington, Ind.	95	6	Madison, Wis.	1 50	9
			Manchester, Vt.	55	4
INDIANAPOLIS, Ind.	1 10	6	Manchester, N. H.	55	4
Iowa City, Iowa	1 75	11	Mansfield, Ohio	75	4
Ithica, N. Y.	65	4	Marengo, Ill.	1 30	8
			Marietta, Ohio	1 10	7
JACINTO, Miss.	1 55	11	Marshall, Mich.	90	5
Jackson, "	2 20	11	Martinsburgh, Va.	75	6
Jackson, Mich.	90	5	Massillon, Ohio	65	4
Jacksonville, Ill.	1 45	9	Maumee City "	70	4
Janesville, Wis.	1 40	9	Maysville, Ky.	1 00	6
Jefferson, N. Y.	65	4	Meadville, Pa.	70	5
Jefferson, Miss.	1 50	8	Medina, N. Y.	40	3
Jefferson, Wis.	1 50	9	Medina, Ohio	75	4
Jeffersonville, Ind.	1 10	8	Memphis, Tenn.	1 75	13
Jersey City, N. J.	20	2	Menassa, Wis.	1 90	10
Jonesville, Mich.	95	5	Meriden, Conn.	20	2
Juliet, Ill.	1 30	8	Michigan City, Ind.	1 00	5
Junction, Ohio	90	6	Middlebury, Vt.	60	4

Charges for Telegraphic Dispatches from New York

List of Offices and Tariff.	Ten words	Each addi'al word.	List of Offices and Tariff.	Ten words	Each addi'al word.
Middleton, Ohio	$1 10	6	North Adams, Mass.	$ 40	3
Midletown, Conn.	20	2	Northampton, "	45	4
Milan, Ohio	60	4	Northfield, Vt.	60	5
Milford, Del.	50	5	Norwalk, Conn.	20	2
Milford, Ohio	1 00	6	Nunda, N. Y.	45	3
Milwaukie, Wis.	1 30	7			
Mineral Point, "	1 50	9	Ogdensburgh, N. Y.	50	3
Misawaukie, Mich.	1 10	6	Orwell, Vt.	60	4
Mobile, Ala.	2 07	12	Oshawa, Can.	75	6
Monroe, Mich.	70	4	Oshkosh, Wis.	1 50	9
Montezuma, Ind.	1 20	6	Oswego, N. Y.	40	3
Montgomery, Ala.	1 85	11	Ottawa, Ill.	1 30	8
Montpelier, Vt.	85	6	Owego, N. Y.	40	3
Montreal, Can.	85	7	Ozaukie, Wis.	1 40	8
Montrose, Pa.	40	3			
Morris, Ill.	1 25	7	Paduca, Ky.	1 57	9
Morrow, Ohio	90	6	Painesville, Ohio	50	3
Mt. Carbon, Pa.	50	4	Palmyra, N. Y.	40	3
Mt. Clements, Mich.	1 00	6	Palmyra, Wis.	1 50	10
Mt. Morris, N. Y.	45	3	Paoli, Ind.	1 40	9
Mt. Pleasant, Ohio	1 30	10	Paris, Ky.	1 35	9
Mt. Sterling, Ill.	1 40	8	Paris, Ill.	1 40	8
Mt. Vernon, Ohio	80	5	Pawtucket, R. I.	20	2
Muscatine, Iowa	1 55	10	Peekskill, N. Y.	20	1
			Penn Yan, "	60	4
Napiersville, Ill.	1 30	7	Peoria, Ind.	1 05	6
Nashville, Tenn.	1 35	10	Peoria, Ill.	1 35	8
Nashua, N. H.	45	4	Perrysville, Ohio	1 10	6
Natches, Miss.	2 25	15	Peru, Ind.	1 00	6
Navarre, Ohio	75	4	Peru, Ill.	1 30	8
Nenah, Wis.	1 90	10	Petersburg, Va.	69	5
Nazareth, Pa.	45	4	Petticodiac, N. B.	1 40	8
Newark, N. J.	20	2	Phoenixville, Pa.	40	4
Newark, Ohio	80	5	Philadelphia, "	25	2
New Albany, Ind.	1 30	8	Picton, N. B.	1 65	11
New Bedford, Mass.	20	2	Piermont, N. Y.	20	1
New Brunswick, N. J.	20	2	Piqua, Ohio	1 30	8
New Buffalo, Mich.	1 00	6	Piketon, "	1 25	8
Newburg, N. Y.	20	1	Pikeville, Ala.	1 65	11
New Castle, Del.	40	4	Pittsburg, Pa.	60	4
New Commerstown, Ohio	70	4	Pittsfield, Mass.	50	4
New Haven, Conn.	20	2	Plattville, Wis.	1 55	10
New Lisbon, Ohio	70	4	Pomeroy, Ohio	1 20	8
New London, Conn.	20	2	Pontiac, Mich.	1 25	8
New Orleans, La.	2 40	14	Port Clinton, Pa.	50	4
New Philadelphia, Ohio	80	4	Port Hope, Can.	75	6
New Richmond, "	1 05	8	Port Huron, Mich.	1 30	8
Newport, R. I.	20	2	Port Jervis, Pa.	25	2
Newton Falls, Ohio	75	4	Portland, Me.	40	3
New Washington, "	80	6	Port Richmond, Pa.	40	3
Niagara, Can.	63	5	Portsmouth, Ohio	1 00	7
Niagara Falls,	45	4	Portsmouth, N. H.	40	3
Niles, Mich.	95	5	Port Washington, Ohio	80	5
Norfolk, Va.	1 19	9	Potosi, Mo.	1 55	10
Norristown, Pa.	40	4	Pottstown, Pa.	40	4

The Electric Telegraph in the United States

LIST OF OFFICES AND TARIFF.	Ten words	Each addit'al word	LIST OF OFFICES AND TARIFF.	Ten words	Each addit'al word
Pottsville, Pa.	$ 50	4	Southport, Wis.	$1 30	8
Poughkeepsie, N. Y.	20	1	Springfield, Mass.	20	2
Prescott, Can.	85	7	Springfield, Vt.	45	4
Princeton, N. J.	20	2	Springfield, Ohio	80	5
Providence, R. I.	20	2	Springfield, Ill.	1 25	8
			Springvalley, Ohio	90	6
QUEBEC, Can.	1 10	9	Stamford, Conn.	20	2
Queenston, "	50	4	Steubenville, Ohio	80	6
Quincy, Ill.	1 30	8	St. Albans, Vt.	60	4
			St. Catharines, Can.	50	4
RACINE, Wis.	1 20	8	St. Charles, Ohio	30	6
Raleigh, N. C.	84	6	St. Clair, Mich.	1 35	8
Randolph, Vt.	50	4	St. Genevieve, Mo.	1 75	10
Ravenna, Ohio	75	4	St. Georges, N. B.	1 15	6
Reading, Pa.	40	4	St. Johns, Can.	1 00	6
Red Hook, N. Y.	20	1	St. Johns, N. B.	1 15	6
Republic, Ohio	80	6	St. Louis, Mo.	1 45	10
Richmond, Pa.	85	3	St. Marys, Ohio	1 00	5
Richmond, Va.	67	5	Sturges Prairie, Mich.	1 00	6
Richmond, Ind.	1 10	6	Suffolk, Va.	1 12	9
Ripley, Ohio	1 05	6	Susquehanna, Pa.	40	2
Rochester, N. Y.	40	3	Syracuse, N. Y.	40	3
Rockford, Ill.	1 35	7			
Rock Island, Ill.	1 55	10	TAUNTON, Mass.	20	2
Rockland, Me.	65	4	Tecumseh, Mich.	90	5
Rocton, N. Y.	30	2	Terrehaute, Ind.	1 10	6
Rocton, Ill.	1 35	7	Theresa, N. Y.	40	3
Rome, N. Y.	30	2	Thomaston, Me.	65	4
Rondout, "	20	1	Thompsonville, Conn.	20	2
Roscoe, Ohio	80	4	Three Rivers, Can.	1 10	9
Rushville, Ill.	1 25	8	Tiffin, Ohio	80	6
Rutland, Vt.	55	4	Toledo, "	70	5
			Tonawanda, N. Y.	50	4
SACKVILLE, N. B.	1 40	8	Toronto, Can.	63	5
Saginaw, Mich.	1 40	8	Trenton, N. J.	20	2
Saco, Me.	40	3	Troy, N. Y.	30	2
Salem, Ill.	1 50	9	Troy, Ohio	80	6
Salem, Ind.	1 00	6	Truro, N. B.	1 65	10
Sandusky, Ohio	60	4	Tuscumbia, Ala.	1 60	12
Saratoga, N. Y.	55	4	Tuscumbia, Me.	1 45	10
Saugerties, "	20	1			
Sauk Prairie, Wis.	2 10	11	UNIONTOWN, PA.	75	5
Savannah, Ga.	1 46	9	Urbana, Ohio	80	6
Schaghticoke, N. Y.	40	3	Urichville, "	80	6
Schenectady, "	30	2	Utica, N. Y.	30	2
Section Ten, Ohio	1 00	6			
Sheboygan, Wis.	1 50	9	VALATIA, N. Y.	20	1
Sheboygan Falls, Wis.	1 50	9	Valparaiso, Ind.	1 10	6
Shelbyville, Ohio	1 05	7	Vergennes, Vt.	60	4
Shelbyville, Ill.	1 05	7	Versailles, Ky.	1 10	7
Shelbyville, Ind.	1 10	6	Vicksburg, Miss.	2 10	15
Shelbyville, Ky.	1 20	8	Vincennes, Ind.	1 10	6
Shullsburg, Wis.	1 60	10			
Sidney, Ohio	1 00	6	WABASH, Ind.	1 00	6
South Bend, Ind.	95	5	Waldoboro, Me.	65	4

Charges for Telegraphic Dispatches from New York

List of Offices and Tariff.	Ten words	Each add'l word.	List of Offices and Tariff.	Ten words	Each add'l word.
Warren, R. I.	$ 20	2	Wilkesbarre, Penn.	$ 55	5
Warren, Ohio	75	4	Williamston, Mass.	40	3
Washington, D. C.	50	5	Wilmington, Del.	45	4
Washington, Ohio	90	6	Winchester, Va.	78	7
Washington, Pa.	85	6	Windsor, Vt.	50	4
Waterbury, Vt.	60	5	Woodfield, Ohio	1 30	9
Watertown, N. Y.	50	3	Woodstock, Vt.	50	5
Waukeesha, Wis.	1 25	7	Woodstock, Can.	88	7
Waynesville, Ohio	90	6	Woodstock, Ill.	1 30	7
Wellsville,	70	4	Woonsocket, R. I.	20	2
Westfield, N. Y.	50	3	Worcester, Mass.	20	2
West Liberty, Ohio	80	6	Wooster, Ohio	75	5
West Randolph, Vt.	50	4			
West Union, Ohio	1 05	8	Xenia, Ohio	90	6
Wheeling, Va.	80	5			
Whitby, Can.	75	6	York, Pa.	50	3
Whitehall, N. Y.	55	4	Youngstown, Pa.	80	5
Whitehaven, Pa.	55	5	Ypsilanti, Mich.	90	5
White River Junction, Vt.	90	6			
Whitewater, Wis.	1 50	7	Zanesville, Ohio	75	5

THE END

If you found this book on the
Electric Telegraph
interesting
you might like
these other titles in
The Electric Telegraph Series

Title:	An Illustrated Handbook to the Electric Telegraph	Title:	The Electric Telegraph - Its History and Progress
Author:	Robert Dodwell	Author:	Edward Highton
Date:	1862	Date:	1852
Edition:	Second	Edition:	First
Pages:	102	Pages:	200
Format:	Paperback	Format:	Paperback
Size:	6 x 9 in (152 x 229 mm)	Size:	6 x 9 in (152 x 229 mm)
ISBN:	978-1-9792525-6-0	ISBN:	978-1-9791199-9-3
Series:	Electric Telegraph 1	Series:	Electric Telegraph 2
Pub:	30 Oct 2017	Pub:	26 Oct 2017

Also available in the
Kindle Store
in e-pub format
as
Kindle Print Book Replicas

Title:	Modern Practice of the Electric Telegraph		Title:	The Electric Telegraph Popularised
Author:	Frank L. Pope		Author:	Dionysius Lardner
Date:	1869		Date:	1867
Edition:	Second		Edition:	Third
Pages:	162		Pages:	346
Format:	Paperback		Format:	Paperback
Size:	6 x 9 in (152 x 229 mm)		Size:	6 x 9 in (152 x 229 mm)
ISBN:	978-1-9818047-1-9		ISBN:	978-1-7963481-4-9
Series:	Electric Telegraph 3		Series:	Electric Telegraph 4
Pub:	18 Dec 2017		Pub:	8 Feb 2019

www.ingramcontent.com/pod-product-compliance
Lightning Source LLC
Chambersburg PA
CBHW051307220526
45468CB00004B/1248